Exoplanets and the Search for Alien Life

Exoplanets and the Search for Alien Life

Nils Caravelli

Mindful Pages

Published in 2025

ISBN: 9789371344425 (PB)
ISBN: 9789371349635 (eBook)

Published by

Mindful Pages
Imprint of Alpha Editions LLC
312 W. 2nd St #1834
Casper, WY 82601, USA

This book was developed using advanced writing tools to support the drafting and ideation process. All content has been thoughtfully curated, refined, and reviewed to ensure quality and relevance.

Content

Beyond the Horizon Why the Search for Alien Life Is Really a Search for Ourselves vii

Chapter 1 Rethinking Life: Beyond the Carbon Bias 1

Chapter 2 The Exoplanet Renaissance: How We Found Thousands of Worlds 26

Chapter 3 Worlds Without Suns: Rogue Planets and Strange Homes for Life 51

Chapter 4 Biosignatures and the Problem of False Positives 74

Chapter 5 AI and the Algorithms of Discovery 98

Chapter 6 The Cosmic Mirror: Psychology and the Need to Know 122

Chapter 7 Alien Contact and the Ethics of Messaging 143

Chapter 8 Religion, Myth, and Alien Revelation 166

Chapter 9 The Future of Exploration: Living Among the Stars 189

Conclusion The Signal and the Silence 211

About the Author 230

Beyond the Horizon
Why the Search for Alien Life Is Really a Search for Ourselves

There is something profoundly human about looking up. Long before we built telescopes or launched satellites, we stared into the night sky and felt both insignificant and infinite. We assigned stories to the stars, tracked their movements, and shaped our calendars around their cycles. But what began as myth-making has now matured into something sharper—an empirical, technology-driven search for truth. The discovery of exoplanets—planets orbiting stars beyond our Sun—has transformed our understanding of the cosmos. Yet this transformation is not merely scientific; it is existential. To study other worlds is, ultimately, to examine our own. This book is about exoplanets, yes—but it is also about identity, imagination, and the boundaries of what we dare to consider as "life."

Astronomy has always occupied a unique position between the abstract and the intimate. It deals with the unimaginable—light-years, galaxies, cosmic origins—while touching on the personal. What does it mean to be alive in such a vast universe? Are we the exception or the rule? These questions, though cast against the backdrop of stars, cut deeply into our psychology. The search for alien life is not only a scientific endeavour; it is also a mirror, reflecting our hopes, fears, and philosophical assumptions. With every detection of a distant world, we are forced to confront our expectations about what life should look like, how it should behave, and whether we would even recognise it if we saw it.

At the heart of this exploration is a paradox. We seek what we do not know, but we are constrained by what we do. Much of our search for alien life has been conducted through an Earth-centric lens—a framework rooted in the only biosphere we've ever studied. Our instruments are tuned to detect atmospheres like ours, temperatures we find hospitable, and molecules essential to our biochemistry. Yet the universe is not obliged to repeat Earth's template. Indeed, to imagine that it would is both statistically unlikely and intellectually timid. If we are to meaningfully search for life beyond Earth, we must first expand our conception of what "life" entails. This book begins with that premise and builds outward.

Chapter 1 begins this reframing by challenging the "carbon bias" that dominates contemporary astrobiology. We critique the limitations of assuming life elsewhere must mirror our own, and delve into the provocative hypothesis of shadow biospheres—forms of life that might exist on Earth, undetected, because they fall outside our conventional definitions. By exploring alternative biochemistries such as silicon-based organisms or ammonia-dependent metabolisms, we begin to understand how deeply our imaginations must stretch. The chapter culminates in a defence of imagination itself, arguing that it is not antithetical to science but essential to its progress. Every major scientific breakthrough began as an act of imaginative courage.

This intellectual flexibility becomes even more critical in Chapter 2, which surveys the rapid acceleration of exoplanet discovery. Thanks to missions like Kepler and TESS, we now know of thousands of worlds orbiting distant stars—many unlike anything in our solar system. The methods used to detect these planets—transit photometry, radial velocity, and spectral analysis—may sound technical, but they are, in essence, acts of cosmic deduction. We read light as though it were a diary. We infer the presence of alien skies from wobbles in starlight. And yet, for all our progress, we remain at the threshold of understanding. Many of the most common planet types—such as super-Earths and mini-Neptunes—have no analogues here. Their atmospheres challenge our models, and their compositions suggest geochemical processes we have yet to theorise. This chapter grounds the reader in the tools of discovery while revealing just how much we still have to learn.

Chapter 3 turns our attention to the unlikeliest of habitats—rogue planets, untethered from any star, drifting through the cold expanse of space. At first glance, they seem inhospitable, even desolate. Yet recent models suggest that some of these worlds may harbour subsurface oceans, warmed by geothermal heat or tidal flexing. If life can adapt to eternal night, it may take on forms we have not imagined: bioluminescent organisms, chemotrophic ecosystems, perhaps even intelligent life adapted to silence and solitude. These speculations are not exercises in fantasy; they are grounded in analogues from our own planet, where extremophiles thrive in places once thought barren. But this chapter also goes deeper. It explores the psychological implications of these lonely worlds—what they symbolise for us, and why their very existence both unsettles and captivates us.

From here, we move into the intricacies of detection in Chapter 4. The concept of a "biosignature"—a chemical marker that suggests biological activity—is central to modern astrobiology. But what counts as evidence? Methane on Mars and oxygen on exoplanets have generated headlines, only to be followed by caveats and contradictions. Nature is deceptive. Many of the markers we associate with life can also be produced abiotically. This raises complex questions about how we interpret ambiguous data, especially when our instruments are limited and our cognitive biases unconsciously influence what we expect to see. This chapter challenges the reader to grapple with uncertainty, ambiguity, and the epistemological limits of science. It proposes that future detection efforts must integrate not just new instruments, but new frameworks—ones that incorporate systems thinking, information theory, and interdisciplinary scepticism.

Chapter 5 continues this theme by examining how artificial intelligence is reshaping the search. Machine learning algorithms now sift through massive datasets, identifying subtle patterns that would elude human analysts. AI has become an indispensable partner in exoplanet research. Yet it also introduces its own challenges. An algorithm is only as imaginative as its training set. If we feed machines our biases, they will replicate them with greater speed and scale. This chapter explores the promise and peril of delegating discovery to non-human intelligences. It raises provocative questions: What if the first to detect alien life is an AI? Would we trust its judgement? Would we even understand its reasoning? And what ethical frameworks should govern autonomous agents that interpret, and possibly act on, signals from other worlds?

At the midpoint of the book, Chapter 6 turns inward. Beneath the science lies psychology. Why are we so obsessed with the idea of alien life? What does it say about us? This chapter explores the emotional architecture of our fascination—from the fear of loneliness to the longing for connection. The Fermi Paradox, often discussed in terms of probability, is recast here as a psychological mirror. Perhaps we worry that the silence of the cosmos reflects something about our own failure to listen—or to survive. The chapter also critiques how we anthropomorphise alien life, projecting our values, behaviours, and even moral frameworks onto the unknown. To truly be open to alien intelligence, we must first confront how deeply human our expectations are.

This introspection leads naturally into Chapter 7, which addresses the ethics of contact. If we were to detect a signal—or send one—who speaks for Earth? What assumptions underlie our messages? Do we presume peace because we value it, or because we lack imagination for the alternative? The chapter dissects ongoing debates about Messaging Extraterrestrial Intelligence (METI), the risks of revealing our location, and the politics of planetary representation. It introduces readers to contact simulations and invites them to consider what they would say, given the chance. Through these exercises, the philosophical dimensions of the search come sharply into focus: discovery is not just about finding others—it is about defining ourselves.

The cultural and spiritual implications of such discovery are explored in Chapter 8. From ancient myths to contemporary religions, humanity has long speculated about beings from the sky. But how would major belief systems respond to confirmed alien life? Could it trigger theological crisis—or catalyse cosmological renewal? This chapter delves into the intersection of religion and science, myth and meaning. It considers the possibility of a "cosmotheology"—a framework that accommodates divinity in a populated universe. The idea of rewriting our origin stories is not metaphorical; it is a literal challenge that could upend foundational narratives about our uniqueness, our purpose, and our place in the cosmos.

Finally, Chapter 9 gazes forward. What happens after detection? What happens if we never find anyone? The future of exploration is not merely about reaching other worlds—it is about deciding what kind of civilisation we want to be. Concepts like interstellar travel, terraforming, and post-biological evolution are discussed not as distant fantasies, but as plausible outcomes within the scope of human ingenuity. Yet with each new frontier comes new responsibility. Colonising an alien world raises profound ethical questions. Encountering a post-biological civilisation forces us to redefine life itself. And venturing far from Earth may mean becoming something other than human. This chapter argues that the search for alien life is also the beginning of our transformation. Even solitude changes us, if we allow it to.

This book does not promise answers. It offers perspectives. It is rooted in science, informed by psychology, sharpened by philosophy, and animated by imagination. Its purpose is not merely to inform, but to expand the reader's conceptual universe. The search for exoplanets is not a detached academic pursuit. It is one of the most profound acts of curiosity a species can undertake. Whether or not we find others, the search itself shapes us. And in the stillness between signals, we discover not only who else might be out there—but who we are.

Chapter 1
Rethinking Life: Beyond the Carbon Bias

For most of human history, life has been understood through a singular, limited lens—our own. Every definition, every model, every assumption has been shaped by a sample size of one: Earth. From the microbial mats of hydrothermal vents to the intricate systems of a rainforest, life on this planet is remarkably diverse, yet fundamentally homogeneous in its chemical composition. It is carbon-based, reliant on liquid water, dependent on DNA or RNA to transmit information, and most often metabolically entwined with oxygen. While this consistency has enabled us to build coherent biological theories and models, it has also tethered our imaginations. In the search for life beyond Earth, this Earth-centric paradigm may be as much a liability as an asset.

This chapter opens with a provocation: What if life exists right now, undetected, not because it is light-years away, but because we have not known how—or where—to look? To answer that question, we must first unravel the assumptions that have guided our search. The so-called "Earth template" has served as a blueprint for astrobiology, influencing the design of our telescopes, the parameters of our detection algorithms, and even the philosophical questions we ask. But as our knowledge of planetary systems expands, and as we confront the sheer variety of exoplanetary environments, it becomes clear that clinging too tightly to Earth's example risks intellectual stagnation. If life elsewhere looks nothing like what we expect, then expecting it to behave like us may cause us to miss it altogether.

This critique is not abstract. The very technologies used to scan distant atmospheres—searching for oxygen, methane, or water vapour—are tailored to environments analogous to our own. Missions to Mars, Europa, or Enceladus often hinge on detecting signs that mimic terrestrial biology. Yet the history of science is filled with examples of paradigm shifts catalysed by those willing to question the prevailing model. In the 19th century, Pasteur's discovery that microbes could cause disease upended centuries of medical orthodoxy. In the mid-20th, the structure of DNA revolutionised genetics by revealing a shared code among living things. We are, perhaps, on the cusp of another such upheaval—one that begins not with new instruments, but with new imagination.

Consider the provocative idea of *shadow biospheres*—a hypothesis that suggests alien life may already exist on Earth, unnoticed because it operates under different biochemical rules. This is not the domain of fringe theory, but a topic that respected scientists such as Carol Cleland and Shelley Copley have explored with rigorous curiosity. A shadow biosphere might consist of organisms based on arsenic instead of phosphorus, or utilise non-standard amino acids, or metabolise in ways that evade our standard staining techniques and culturing protocols. In 2010, a NASA-funded team led by Felisa Wolfe-Simon claimed to have discovered bacteria in California's Mono Lake that could incorporate arsenic into their DNA. Although the findings were later disputed, the controversy underscored the central issue: our tools, shaped by conventional biochemistry, may be blind to truly alien forms of life. The absence of evidence is not evidence of absence—it may be evidence of narrow methodology.

This brings us to the broader question of alternative biochemistries. Why assume carbon is the only viable backbone for life? Carbon's versatility in forming complex, stable molecules makes it an excellent candidate, but not the exclusive one. Silicon, for instance, shares chemical similarities with carbon and is abundant in the universe. While silicon-based life has often been relegated to science fiction, theoretical models suggest that under certain conditions—such as high temperatures or in solvent systems other than water—silicon could form the basis for self-replicating, information-bearing molecules. The late Isaac Asimov speculated on such life forms in his essays, and while no empirical evidence yet supports their existence, the thought experiment forces a critical re-evaluation of our biological assumptions.

Ammonia is another candidate deserving attention. Though toxic to most Earth life, ammonia has properties similar to water—it can dissolve organic molecules, has a relatively high boiling point under pressure, and could serve as a solvent in extremely cold environments. Saturn's moon Titan, for example, possesses lakes of liquid methane and a thick atmosphere rich in organic compounds. Here, life—if it exists—would not merely be alien in form, but in function. It might metabolise using hydrogen, exhale ethane, and reproduce through entirely unfamiliar pathways. The implications for detection are profound. Our current instruments are calibrated to detect oxygen-rich, water-bearing planets because that is what supports us. But an ammonia-based ecosystem on Titan would leave behind a very different set of biosignatures. Without expanding our biochemical imagination, we risk sending elegant spacecraft to alien environments equipped to discover only ourselves.

Equally revolutionary is the notion of life without oxygen. On Earth, aerobic respiration dominates, but it is not universal. Anaerobic microbes thrive in environments devoid of oxygen, from the deep ocean to the guts of animals. These organisms metabolise using sulphates, nitrates, or even iron. In methane-rich environments, methanogens convert carbon dioxide and hydrogen into methane, sustaining themselves through a biochemistry that would be considered toxic to most life. The presence of such extremophiles on Earth demonstrates that biology is not inherently wedded to oxygen—or even to sunlight. Jupiter's moon Europa and Saturn's moon Enceladus both possess subsurface oceans kept liquid by tidal heating, potentially providing environments for chemosynthetic life. Meanwhile, some exoplanets orbiting red dwarfs may never receive the full spectrum of visible light. If life there exists, it would have evolved mechanisms of energy conversion entirely unlike photosynthesis. Such possibilities expand the canvas on which we paint our theories of habitability.

Yet as we broaden our view, we encounter a curious tension: scientific inquiry thrives on rigour, but discovery often begins with wonder. The role of imagination in science is too often dismissed, as though hypothesising beyond the data were a form of intellectual betrayal. But every great leap in understanding has required the audacity to think beyond evidence. Galileo imagined a Sun-centred solar system before he could fully prove it. Darwin envisioned a tree of life before molecular biology confirmed it.

Einstein conceived of spacetime curvature before gravitational waves were observed. Imagination, when disciplined by logic and informed by observation, becomes a tool of scientific advancement—not a threat to it.

The search for life beyond Earth demands just such a fusion. It requires us to move beyond the constraints of the known, to model life not only as it is, but as it *could be.* This means engaging with interdisciplinary perspectives—from chemistry and physics to philosophy and speculative fiction. It means cultivating a tolerance for uncertainty and a reverence for possibility. If we are to detect truly alien life, we must be willing to entertain alien thinking. This begins not with algorithms or apparatus, but with a willingness to ask better questions.

What if life does not require DNA? What if it arises not from carbon, but from a compound yet unknown to us? What if intelligence does not produce technology, or if its signals are so different from ours that we mistake them for noise? What if the cosmos is already teeming with life, but we've defined it out of existence through our narrow metrics? These are not idle musings; they are the intellectual scaffolding of a new astrobiology—one that is less about searching for Earth 2.0 and more about recognising life in all its unfathomable diversity.

This chapter lays the philosophical and empirical groundwork for that shift. It argues that expanding our definition of life is not a luxury, but a necessity. It invites the reader to consider that our current ignorance is not a failing, but a frontier. And it challenges the scientific community to embrace imagination not as a soft science, but as a hard prerequisite for progress.

To rethink life is to rethink ourselves. To search for the alien is to confront the limitations of the familiar. And to go beyond the carbon bias is not merely to look outward—it is to awaken a deeper capacity for wonder, rigour, and humility in the face of the unknown. As the rest of this book will show, the cosmos may not resemble us. But it may still *include* us—if we are willing to see with more than just Earth-trained eyes.

The Earth Template: A Flawed Standard

For all the ambition embedded in our search for extraterrestrial life, it is striking how narrowly defined our expectations often remain. Across decades of research, exploration, and hypothesis, we have largely pursued life as we know it—life that resembles Earth's biosphere in chemical structure, environmental requirements, and metabolic processes. While this Earth-centric approach has provided a practical starting point for scientific investigation, it is also a conceptual constraint, one that risks obscuring forms of life that diverge significantly from our own. As our instruments grow more powerful and our catalogues of exoplanets expand, the question is no longer whether we are alone, but whether we are even looking in the right way.

The assumption that alien life must resemble terrestrial life is not without merit. After all, Earth is the only known example of a life-bearing planet, and science is by nature inductive—it extrapolates from the data at hand. Carbon, water, oxygen, DNA: these are the materials and mechanisms we understand. It makes sense to build models around them, to engineer instruments that search for familiar signatures in distant atmospheres. Yet the very familiarity of these elements can blind us. By centring our detection strategies on what we already know, we risk missing what we do not—even if it is right in front of us.

This bias is especially evident in the design of space missions. The Mars rovers, for example, have been tasked with seeking out signs of water and organic molecules, under the premise that where there is water, there may be life. But this presumption is rooted in the Earth template—on our planet, water is a universal solvent and a fundamental component of all known life. However, the presence of water does not guarantee habitability elsewhere, nor does its absence imply lifelessness. Life might have evolved in solvents such as ammonia or methane, or used chemical processes that do not require liquid media at all. By using Earth as the gold standard, we unintentionally narrow the scope of our inquiry, rendering the unfamiliar invisible.

This limitation is not only technological but philosophical. It reflects a deeper epistemological issue: our difficulty in imagining what we have not experienced. The anthropologist Edmund Leach once remarked that "the limits of our language are the limits of our world." In astrobiology, this idea takes on new urgency. Our definitions of "life" and "habitability" are embedded in linguistic and cultural assumptions shaped by our specific evolutionary and planetary context. These definitions serve as both map and blindfold. They guide us, but they also filter out that which does not conform.

Real-world examples help to illuminate this tension. Consider the controversial 2010 announcement of bacteria from Mono Lake, California, that appeared to incorporate arsenic into their DNA instead of phosphorus. The research, led by Felisa Wolfe-Simon and funded by NASA's Astrobiology Institute, ignited a firestorm of debate. Critics questioned the methods and reproducibility of the findings, and subsequent studies cast doubt on whether arsenic had truly been substituted in the DNA backbone. Yet the deeper significance of the episode lies not in whether the specific claim was validated, but in the reaction it provoked. The scientific community was compelled to confront the fragility of its assumptions. Could life, even on Earth, exist outside the biochemical boundaries we consider universal? And if it could here, what might be possible elsewhere?

The concept of a "shadow biosphere" arises directly from this line of inquiry. Proposed by scientists such as Carol Cleland and Shelley Copley, the idea posits that a second, undetected form of life may exist on Earth—microbial ecosystems that do not interact with conventional biochemistry and therefore elude standard methods of detection. These hypothetical organisms might metabolise differently, store genetic information in alternative molecules, or depend on elements like arsenic or silicon. If such life exists, it would not only challenge our definitions but also indicate that the universe may be far more hospitable to life than current estimates suggest.

Despite its compelling implications, the Earth template persists as a dominant framework in both scientific and popular discourse. This is partly due to practicality: designing instruments to detect every conceivable form of life is logistically and financially impossible. It is also a function of institutional inertia. Scientific paradigms, as Thomas Kuhn famously argued, tend to resist change until anomalies accumulate to the point of rupture. We have not yet reached that rupture, but we are certainly accumulating anomalies—planets that do not conform to our models, atmospheres with inexplicable compositions, extremophiles that thrive in boiling acid or crushing pressure. These data points are not just curiosities; they are clues, fragments of a larger picture we have yet to assemble.

One of the most striking examples comes from the discovery of extremophiles on Earth—organisms that flourish in conditions previously considered uninhabitable. In the depths of hydrothermal vents, in subglacial Antarctic lakes, and within the radioactive ruins of Chernobyl, life not only persists but adapts. These organisms metabolise without sunlight, survive freezing or boiling temperatures, and tolerate extreme acidity or salinity. Their existence forces a reconsideration of what constitutes a habitable environment. If life can adapt so radically on Earth, it stands to reason that it could arise in equally unfamiliar conditions elsewhere.

Yet habitability is not just about environmental conditions—it is also about chemical plausibility. Carbon, for example, is the backbone of terrestrial life due to its ability to form stable, complex chains. But silicon, its chemical cousin, also possesses this capacity under certain conditions. Theoretically, silicon-based life could emerge in high-temperature environments where silicon compounds are more stable than their carbon equivalents. While such organisms remain speculative, their plausibility is sufficient to merit consideration. Dismissing them outright because they do not resemble what we know is not science—it is dogma.

The flawed standard of the Earth template also impacts how we interpret ambiguous data. When scientists detect an unusual atmospheric signature on an exoplanet—say, an excess of methane or oxygen—it is tempting to jump to conclusions based on terrestrial analogues. But as many researchers have pointed out, these molecules can also be produced through non-biological processes. Interpreting them through an Earth-centric framework risks both false positives and false negatives. A truly alien biosphere may produce biosignatures that we do not currently associate with life—or it may produce no recognisable signal at all.

To address this, some researchers advocate for a more agnostic approach to life detection—one that focuses not on specific molecules, but on patterns of disequilibrium, entropy reduction, or information processing. These signatures, while more abstract, may prove more universal. They suggest a broader definition of life: not as a specific chemical arrangement, but as a system capable of self-organisation, replication, and adaptive complexity. Adopting this perspective requires a conceptual shift, but it also opens up vast new territories for exploration.

This shift is already underway in some quarters. NASA's Origins, Spectral Interpretation, Resource Identification, Security, Regolith Explorer (OSIRIS-REx) mission and the forthcoming James Webb Space Telescope reflect a growing commitment to multi-spectral, high-resolution data collection capable of detecting a wider range of planetary phenomena. Meanwhile, synthetic biology and artificial life research are beginning to create life-like systems from non-traditional chemistries, further challenging the idea that life must look a certain way. As our scientific toolkit expands, so too must our willingness to question foundational assumptions.

Ultimately, the Earth template is not inherently flawed—it is just incomplete. Like any model, it simplifies complexity in order to be useful. But models must evolve as new evidence accumulates, and our evidence now points to a cosmos that is stranger, more diverse, and more surprising than we once imagined. The standard we apply to life must be flexible enough to accommodate that strangeness. It must be informed by what we know, but not imprisoned by it.

To move beyond the Earth template is not to abandon rigour—it is to embrace intellectual humility. It is to recognise that our understanding of life is provisional, and that the universe may hold examples that defy every category we have constructed. This is not a reason for despair; it is an invitation to expand our inquiry, to sharpen our tools, and to rediscover the spirit of wonder that lies at the heart of all scientific endeavour.

The more we search, the more we must be willing to see. Not just what we expect, but what we cannot yet imagine. For it may be that the alien is not distant, but simply different—and that the first step toward discovery is learning how to recognise difference not as deviation, but as potential.

Shadow Biospheres: Hidden Life, Even Here

For all our cosmic ambition and technological sophistication, there is an uncomfortable possibility we must confront: alien life may already be here—and we may have failed to recognise it. Not hidden among the stars or cloaked in distant atmospheres, but potentially present on our own planet. This is the provocative foundation of the shadow biosphere hypothesis, a theory that suggests Earth may harbour a parallel form of life so fundamentally different in biochemistry that it escapes our detection entirely. If true, it would not only revolutionise our understanding of biology but also upend the very premise of the search for extraterrestrial life, shifting it from a question of distance to one of definition.

The notion of a shadow biosphere challenges a central tenet of modern biology: that all life on Earth shares a common ancestry and a uniform biochemistry. This assumption has served us well for centuries. The discovery of DNA as the universal genetic code, the centrality of carbon as life's molecular scaffold, and the ubiquitous role of water as solvent have formed a coherent framework for understanding life's origins and mechanisms. Yet, as with all scientific paradigms, coherence should not be mistaken for completeness. The shadow biosphere hypothesis emerges not from a rejection of existing biology but from its logical extension: if life could have emerged multiple times under different conditions, why should all examples of it follow the same rules?

The term "shadow biosphere" was popularised by astrobiologist Carol Cleland and microbiologist Shelley Copley at the University of Colorado, who proposed that Earth may host life forms with radically different molecular structures—perhaps based on alternative amino acids, nucleic acids, or elemental compositions. The reason such life has remained undetected, they argue, is that our standard methods for identifying microorganisms are inherently biased toward organisms that resemble known life. Microbial life is typically detected through culturing techniques, DNA sequencing, or metabolic assays, all of which presume particular chemical and structural features. A life form with a fundamentally distinct biochemistry would not grow on standard culture media, might not bind to common stains, and could be entirely overlooked during genomic analysis if its genetic code operates outside the known paradigm.

A striking case that brought this hypothesis into public focus emerged in 2010, when a team led by Felisa Wolfe-Simon reported the discovery of bacteria in Mono Lake, California, capable of incorporating arsenic into their DNA in place of phosphorus. The announcement, supported by NASA's Astrobiology Institute, generated international headlines and an immediate scientific uproar. Phosphorus is one of life's essential elements, forming the backbone of DNA and playing a critical role in energy transfer via ATP. The possibility that arsenic—typically toxic to living organisms—could substitute for phosphorus suggested that biochemistry was more flexible than previously assumed. Critics swiftly challenged the methodology and interpretation of the findings, with follow-up studies casting doubt on the extent of arsenic incorporation. Yet irrespective of the specific outcome, the episode served a valuable function: it exposed how brittle our biological assumptions can be, and how easily our methods might overlook alternative life forms.

The search for shadow life on Earth continues, albeit quietly, in fringe laboratories and interdisciplinary collaborations. One promising line of inquiry lies in deserts, deep-sea hydrothermal vents, and hypersaline environments—regions where extreme conditions might favour or preserve unconventional biochemistries. Such environments are already known to host extremophiles: microbes that survive in boiling acid, thrive in radioactive waste, or metabolise metals in the absence of light. While extremophiles still conform to conventional carbon-based life, their resilience hints at biology's latent versatility. If life can stretch so far within the known bounds, perhaps it also exists outside them altogether.

The implications of such a discovery would be profound. First, it would force a redefinition of life itself. At present, life is typically described by a list of characteristics: metabolism, replication, evolution, and so forth. But these are properties derived from Earth life, not necessary features of all possible biological systems. A shadow biosphere would suggest that life is a broader phenomenon than we currently conceive—a set of principles that can manifest in multiple, unrelated chemistries. This could radically expand the types of environments deemed habitable, both on Earth and beyond. Worlds previously dismissed as sterile due to lack of water or the presence of toxic elements might instead be fertile grounds for life, if only we knew how to look.

Second, it would recalibrate our search for alien life across the cosmos. Much of astrobiology today is shaped by what we might call "biochemical presumption." Space missions search for signs of water, oxygen, and organic carbon, assuming that alien life will be similar to us. But if alternative life has already evolved independently on Earth, then it is reasonable to assume it could emerge under equally diverse conditions elsewhere. Such a revelation would push space agencies to design new detection tools—ones that do not rely on Earth-centric markers, but instead search for broader indicators of complexity, non-equilibrium systems, or anomalous energy flows that might signal biological processes.

Moreover, the discovery of a shadow biosphere would reframe the origin-of-life debate. Current hypotheses about abiogenesis—the process by which life arose from non-living matter—are constrained by the assumption that life emerged only once, in a singular event leading to the last universal common ancestor (LUCA). But if multiple origins are possible, then life may be a far more common outcome of planetary chemistry than we have assumed. Perhaps the emergence of life is not a rare accident, but a frequent response to a certain threshold of complexity. The cosmos could be not just biologically active, but biologically prolific.

There are also philosophical and psychological ramifications to consider. The idea that alien life might exist not in the stars, but in our own biosphere, unsettles our sense of distance and otherness. It suggests that we are not the sole standard-bearers of life on this planet, but participants in a more diverse and mysterious ecological reality. Such a realisation might erode anthropocentric thinking and promote a deeper sense of humility in our scientific and ethical dealings with the natural world. If there are forms of life on Earth we do not yet understand, how can we claim certainty about life elsewhere?

The pursuit of shadow biospheres is not without its challenges. Technologically, we lack tools fine-tuned for detecting unfamiliar biochemistries. Most microbiological assays assume the presence of ribosomal RNA or universal proteins—elements that would not exist in an alien lineage. Even basic assumptions about cell structure or membrane permeability may not apply. Conceptually, the field suffers from a paucity of testable predictions. It is difficult to search for something when one cannot define what one is looking for. This has led some researchers to dismiss the hypothesis as speculative, even metaphysical. Yet such criticisms echo those once levelled at the concept of extremophiles, which were once deemed improbable before being discovered in abundance. Science progresses not by clinging to what is known, but by testing the boundaries of possibility.

Encouragingly, new technologies are beginning to close the gap. Advances in non-targeted metabolomics, high-throughput microscopy, and machine learning are enabling researchers to analyse environmental samples without presupposing specific molecular structures. Synthetic biology, meanwhile, is beginning to craft life-like systems with non-standard nucleotides and alternative amino acid sets, offering insights into how different life might plausibly function. As our tools become less biased and more inclusive, the possibility of detecting a hidden biosphere becomes increasingly realistic.

The concept of a shadow biosphere reminds us that discovery is not always a matter of distance. Sometimes, the most profound revelations lie hidden in the data we overlook, the patterns we dismiss, or the hypotheses we fail to entertain. To search for alien life is to ask whether biology is universal or parochial—whether it is shaped by necessity or by contingency. The shadow biosphere suggests an answer that is both thrilling and unsettling: that life may not be the exception, but the rule; not singular, but plural; not out there, but already here.

If this is true, then we are not merely observers of a biological universe—we are surrounded by it, immersed in it, and perhaps even blind to much of its wonder. To see what has long been hidden, we must do more than refine our instruments. We must refine our expectations. And in doing so, we may finally begin to understand not just life as we know it, but life as it could be.

Silicon, Ammonia, and Beyond: Alternative Biochemistries

What if life exists in forms we cannot currently recognise—organisms built not on carbon and water, but on silicon and ammonia, or perhaps even on elements and structures unknown to present science? While such possibilities may seem like the domain of speculative fiction, they are increasingly the subject of serious scientific inquiry. As we move further into the era of exoplanet discovery, the question of what life might look like beyond Earth grows more urgent—and more complex. The vast diversity of planetary environments observed in recent years demands a corresponding flexibility in our biological imagination. Yet our detection technologies and conceptual frameworks remain anchored to a single model: the Earth template. To break free of this limitation, we must consider what alternative biochemistries could look like and what they mean for our search for life.

At the core of this inquiry is the recognition that life, as it exists on Earth, is carbon-based. Carbon's unparalleled ability to form stable, complex, and diverse molecular structures has made it the foundation of terrestrial biology. Its unique capacity to create four covalent bonds allows for the intricate architecture of proteins, nucleic acids, carbohydrates, and lipids—the chemical basis of life as we know it. Water, similarly, serves as the universal solvent, providing a medium for biochemical reactions and regulating thermal stability. Yet the question remains: are carbon and water essential for life anywhere, or are they simply the ingredients available to Earth?

To answer this, scientists have turned to silicon as a plausible alternative to carbon. Chemically, silicon resides just below carbon in the periodic table and shares several key properties, including the ability to form four covalent bonds. This suggests that, under the right conditions, silicon could support complex molecular structures. The concept of silicon-based life has captured the scientific imagination for decades, appearing in both theoretical papers and popular culture. However, silicon comes with limitations. Unlike carbon, it tends to form rigid, less versatile chains, and its compounds—such as silanes—are generally less stable in the presence of water. Silicon dioxide, the analogue of carbon dioxide, is a solid at Earth-like temperatures, which could pose problems for respiration and excretion in a silicon-based organism.

Nonetheless, under different environmental conditions—particularly at high temperatures or in non-aqueous solvents—these limitations may not apply. Astrobiologists have speculated that in environments such as the superheated atmospheres of certain exoplanets, or the crusts of geologically active moons, silicon-based biochemistry could be not only viable but optimal. For instance, a planet with a sulphur-rich atmosphere or abundant hydrogen fluoride might support silicon compounds in liquid form, enabling dynamic chemistry alien to our own. Some theoretical models propose that high-pressure environments could further stabilise silicon chains, allowing for the development of complex, self-replicating systems.

If silicon represents one axis of alternative biochemistry, ammonia is another. Often viewed as a potential substitute for water, ammonia has several properties that make it a credible candidate as a biological solvent. It remains liquid at temperatures far below water's freezing point, has a high dielectric constant, and can dissolve many organic and inorganic compounds. Ammonia-based life could conceivably thrive in environments like Saturn's moon Titan, which features lakes of liquid methane and ethane on its surface and is believed to contain subsurface ammonia-water oceans. These conditions are hostile to Earth life, but for organisms adapted to such chemistry, they could be ideal.

One of the most compelling aspects of ammonia is its ability to support acid-base reactions and hydrogen bonding—critical mechanisms in biological processes. Its lower surface tension and higher volatility may allow for faster diffusion rates and different cellular dynamics. Life based in ammonia would likely operate at much colder temperatures and involve slower metabolic rates, but this does not preclude complexity. In fact, some scientists posit that ammonia-based organisms could evolve along very different biological trajectories, developing biochemical systems adapted for energy extraction and replication in conditions that would freeze terrestrial life solid.

These alternative chemistries force us to rethink not only the environments we deem habitable, but also the instruments we use to detect life. Current missions searching for biosignatures—such as oxygen, methane, or complex organic molecules—are based on Earth-centric assumptions. They rely on spectrometers tuned to detect the light-absorbing properties of molecules familiar to us. But silicon-based life may exhale silane instead of carbon dioxide. Ammonia-based metabolisms may generate waste products entirely foreign to our catalogues. If our tools are not designed to recognise these markers, we may overlook clear evidence of alien life simply because it fails to fit our expectations.

This challenge is not merely theoretical. The detection of exoplanet atmospheres has already revealed chemical profiles that defy conventional models. Planets such as WASP-121b and HD 209458b exhibit atmospheric constituents—like metal oxides and unexpected cloud formations—that point to a bewildering diversity of planetary chemistry. As we refine our spectroscopic techniques, we are beginning to see that nature may be far more inventive than we had anticipated. Yet the question remains: will we know alien life when we see it?

One approach gaining traction involves shifting from molecular specificity to systemic patterns. Rather than search for known biosignatures, scientists are beginning to look for thermodynamic and information-theoretic anomalies—situations where entropy is locally reversed, or where energy flows and chemical distributions deviate from what is expected in abiotic systems. These are more universal indicators of life-like processes, irrespective of the underlying chemistry. For example, life—regardless of its molecular makeup—tends to produce chemical disequilibrium. It maintains order against entropy, often in ways that are observable across planetary scales.

Another promising direction comes from synthetic biology. By creating artificial life forms in the laboratory—organisms with non-standard amino acids, artificial bases in their DNA, or even entirely synthetic cells—researchers are exploring the boundaries of what constitutes a viable living system. These engineered organisms serve as test cases for how alternative biochemistries might function and evolve. NASA's Ames Research Center and DARPA have both funded projects in xenobiology, an emerging field dedicated to the study of life forms with "alien" chemistries. Such work not only prepares us for discovering unfamiliar life forms but also helps ensure our detection technologies do not miss them.

Historically, the tendency to universalise Earth-based life is understandable. It is a pragmatic starting point. But as our observational reach extends beyond our solar system, clinging to a single biochemical model becomes increasingly untenable. Already, the cataloguing of exoplanets has revealed systems with wildly diverse properties—planets with iron rain, diamond cores, and atmospheres of hydrogen sulphide. It is illogical to assume that life, if it exists on these worlds, would adhere to a blueprint derived solely from Earth.

The broader philosophical implication is that life may be a process rather than a substance—defined not by specific molecules, but by what it *does*: replicate, adapt, metabolise, evolve. If so, our definitions must become more inclusive, and our technologies more versatile. We must be willing to detect the unexpected, to design instruments capable of reading signatures not yet imagined, and to interpret data without the filter of familiar life. This is not merely a scientific challenge, but a cognitive one. It requires a recalibration of our curiosity, a shift from looking for another Earth to seeking evidence of complexity in whatever form it may take.

Such an epistemological shift also carries ethical weight. If we do discover life—especially one based on unfamiliar chemistry—we will be forced to confront questions about its rights, its value, and our obligations toward it. Will a silicon-based microbial colony on an exoplanet be deemed worthy of protection? How will we respond to a biosphere that functions according to principles we do not fully understand? The recognition of alternative life forms may challenge not only our scientific models, but our moral frameworks as well.

Ultimately, the exploration of alternative biochemistries is not about discarding what we know, but about enlarging what we can know. It is about embracing complexity, uncertainty, and the rich potential of the unknown. In doing so, we open ourselves to a cosmos that is not only more diverse than we imagined, but more alive. And perhaps, in expanding our definition of life, we also expand our understanding of ourselves—not as the standard, but as one instance of nature's astonishing capacity for invention.

Metabolism Without Oxygen: Alien Energy Systems

When contemplating the possibility of alien life, one of the most pervasive—and limiting—assumptions we bring is that oxygen is essential to metabolism. On Earth, aerobic respiration dominates complex life; it is an efficient system that enables organisms to convert glucose into energy with high yields. Oxygen, in this context, is not just a gas—it is a cornerstone of our understanding of vitality. Yet this view, while reflective of Earth's evolutionary trajectory, is parochial. As astrobiology broadens its scope beyond our biosphere, the question is not whether life can exist without oxygen, but how it might flourish in environments governed by entirely different energy systems. The adaptability of metabolism, as it turns out, may be one of the most profound indicators of life's potential universality.

Even on Earth, oxygen is not the baseline for all biology. The earliest forms of life emerged in anoxic conditions, long before photosynthesis began to oxygenate the atmosphere. These ancient organisms relied on anaerobic respiration—metabolic pathways that extract energy without using oxygen. Some bacteria use sulphates, nitrates, iron, or carbon dioxide as terminal electron acceptors. Others, such as methanogens, produce methane as a metabolic byproduct by reducing carbon dioxide with hydrogen. These extremophiles are not mere relics of a distant past; they thrive today in oxygen-deprived environments like deep-sea hydrothermal vents, underground aquifers, and even the digestive tracts of ruminants. Their existence not only confirms that metabolism is possible without oxygen, but also suggests that life elsewhere in the universe may operate on very different biochemical foundations.

The significance of such life forms becomes even more pronounced when we turn our gaze outward to worlds like Saturn's moon Titan or Jupiter's moon Europa. Titan, with its thick nitrogen atmosphere and surface lakes of methane and ethane, presents a scenario radically different from Earth. Its average surface temperature hovers around -179°C, far below water's freezing point. Yet beneath this alien veneer lies a tantalising possibility: that some form of metabolism could occur using methane as a solvent and hydrogen as a fuel. In 2010, NASA scientists examining data from the Cassini-Huygens mission noted unusual patterns in the distribution of hydrogen and acetylene near Titan's surface. One interpretation of this anomaly suggested the possible consumption of hydrogen by hypothetical organisms—a biological process, if real, that would operate entirely without oxygen and outside the framework of water-based chemistry.

Such speculation, while cautious, is grounded in real biochemical logic. Just as Earth's methanogens consume hydrogen and exhale methane, it is conceivable that Titanian life could reverse the equation—metabolising hydrogen in a methane-rich environment. This form of life would not merely tolerate the cold and chemical extremity; it would be shaped by it, evolving biochemical adaptations that exploit rather than resist its surroundings. The metabolic pathways would be unfamiliar but theoretically sound, involving reactions driven by redox gradients not reliant on oxygen but on more abundant molecules like hydrogen or ethane.

Europa, by contrast, offers a different but equally compelling scenario. Beneath its icy crust lies a global subsurface ocean, likely kept liquid by tidal flexing from Jupiter's immense gravitational pull. This hidden sea may contain more than twice the volume of all Earth's oceans combined. What makes Europa particularly intriguing is the potential for hydrothermal activity at the ocean floor. On Earth, such vents host entire ecosystems powered not by sunlight, but by chemosynthesis. In these environments, bacteria oxidise hydrogen sulphide or methane to produce energy, forming the base of a food chain that includes giant tube worms, crustaceans, and other specialised life forms.

If similar conditions exist on Europa, there is no reason—biologically speaking—that comparable life forms could not emerge. The vent hypothesis for Europa does not rely on oxygen but on the availability of chemical disequilibrium—a condition where reactants and products do not exist in thermodynamic balance, creating the potential for energy extraction. This is a key insight: life does not require specific substances so much as it requires a mechanism for capturing and storing energy in a stable, replicable way. Whether that energy comes from oxidising glucose with oxygen or reacting hydrogen with carbon dioxide is a matter of local chemistry, not biological necessity.

This redefinition of metabolism has important implications for how we search for alien life. Current life-detection instruments often rely on biomarkers tailored to Earth-like conditions—oxygen, ozone, methane in the context of a nitrogen-oxygen atmosphere. But in worlds where oxygen is absent, such markers are not only unhelpful but misleading. A more agnostic approach is needed—one that focuses on identifying environments with persistent energy gradients, catalytic interfaces, and sources of molecular complexity. Spectrometers must be designed not only to detect Earth's metabolic byproducts but to scan for broader chemical signatures of disequilibrium. The future of life detection lies not in searching for specific molecules but in identifying systemic anomalies—patterns that suggest active chemistry shaped by evolutionary processes.

There is also a compelling analogy in artificial life research. In laboratory settings, scientists have created protocells and metabolic networks that do not depend on oxygen. Some synthetic systems utilise redox reactions involving metal ions or use ATP analogues derived from non-phosphate sources. These experiments demonstrate that the logic of metabolism is broader than the specific components employed by nature on Earth. They provide working examples of alternative energy systems that could feasibly emerge under different planetary conditions. The more we tinker with life in the lab, the more we realise how adaptable—and improvable—biology may be.

One of the more radical possibilities emerging from this discourse is that intelligent life might also arise in non-oxygenated environments. Traditionally, the development of intelligence has been linked to high metabolic efficiency, something oxygen provides. Aerobic respiration yields significantly more energy per glucose molecule than anaerobic processes, enabling the development of energy-intensive organs like the brain. But this is an Earth-bound assumption. It is entirely possible that different forms of intelligence could evolve with lower energy budgets, through slower metabolism, distributed cognition, or even biochemical mechanisms we have yet to conceive. Intelligence, like life itself, may be more a function of complexity and adaptability than of specific molecular prerequisites.

Such reflections demand not only scientific humility but a philosophical reframing. We must acknowledge that Earth's biosphere is just one solution space within a much larger problem set. Evolution does not seek perfection; it seeks sufficiency within context. The biochemistry of any life form is likely a function of its environment's constraints and affordances—not a universal design but a local adaptation. By assuming oxygen as a prerequisite, we are not just narrowing our field of view—we are imposing our planetary parochialism on a cosmos that may operate by very different rules.

As our search for life continues, from the subsurface oceans of Enceladus to the hydrogen-rich atmospheres of mini-Neptunes, the lesson is clear: metabolism is a process, not a prescription. It is the art of survival within energy gradients, regardless of the molecules involved. Life, if it exists beyond Earth, will not ask whether we are prepared to recognise it. It will simply function, flourish, and perhaps even think—according to its own thermodynamic rules.

To meet that challenge, we must broaden our criteria, refine our instruments, and above all, expand our imagination. Biology is not static; it is an open-ended system shaped by the constraints of its chemistry and the opportunities of its environment. If the cosmos harbours life beyond our oxygenated shores, it will not wait for our approval. It will simply *be*. And our readiness to discover it will depend not on how advanced our technology is, but on how willing we are to redefine what we mean by "alive."

The Role of Imagination in Science

In the popular imagination, science is often portrayed as the solemn realm of facts, formulas, and rigorously controlled experiments. It is regarded as the antithesis of fantasy—a methodical, grounded pursuit of truth that stands apart from the fluidity of dreams and speculation. Yet this narrative conceals a deeper and more nuanced reality. Far from being a discipline of mere deduction, science thrives on creativity. Every breakthrough begins not with data, but with a question that has not yet been asked. And behind each question lies the faculty that dares to conceive of the unknown: imagination.

Nowhere is this more evident than in the field of astrobiology, where the search for life beyond Earth hinges on our ability to envision what we have never seen. Here, imagination is not ancillary to science—it is indispensable. The hypothesis that life may exist on rogue planets, in ammonia seas, or within methane clouds requires imaginative leaps that traditional frameworks may find difficult to accommodate. Yet without these mental excursions, our investigations remain trapped within the narrow boundaries of the familiar.

History repeatedly affirms the catalytic power of imagination in scientific advancement. In 1610, Galileo Galilei turned his telescope toward Jupiter and observed its moons orbiting the planet, directly challenging the geocentric model of the cosmos. This was not a mere empirical act—it was guided by a willingness to reimagine the structure of the universe. In the 19th century, Dmitri Mendeleev's formulation of the periodic table was not simply a collation of known elements, but a conceptual structure with gaps that predicted the existence of elements yet to be discovered. These gaps were filled years later by the likes of gallium and germanium, validating not only the table's logic but Mendeleev's ability to think beyond the available evidence.

Theoretical physics is perhaps the most striking testament to this principle. Albert Einstein's theory of general relativity required imagining spacetime as a malleable fabric, contorted by mass and energy—an idea with no immediate observational support at the time. It was an act of mental modelling so profound that it took decades for empirical technologies to catch up. The discovery of gravitational waves a century later vindicated his theoretical vision and reminded us that science is not a list of facts, but a living process of re-conceptualisation.

Within biology, too, major shifts have relied on imaginative frameworks. The double helix structure of DNA, revealed through the combined insights of Watson, Crick, Franklin, and Wilkins, emerged from the imaginative synthesis of X-ray diffraction data and chemical intuition. Similarly, Lynn Margulis's endosymbiotic theory—that mitochondria and chloroplasts originated as independent organisms that merged with a host cell—was initially dismissed as speculative fiction. Today, it forms a foundational element of evolutionary biology. In each case, the crucial ingredient was not merely empirical data, but the capacity to interpret that data through a lens that had not previously existed.

When it comes to exoplanet research, the need for imaginative thinking becomes even more acute. We are now discovering planets unlike anything in our solar system—hot Jupiters that skim their stars in tight orbits, super-Earths with crushing gravity, and ice giants with exotic atmospheres. To make sense of these worlds, and to assess their potential for hosting life, we must resist the temptation to extrapolate directly from Earth. Our planet may be the only template we have, but it cannot be the only one we trust. Instead, we must learn to ask: What would life look like if the rules were different? What forms might biology take in conditions radically unlike our own?

This is not idle speculation—it is a methodological necessity. Consider Titan, Saturn's largest moon, with its methane lakes and hydrocarbon haze. If life exists there, it would likely be non-aqueous and operate at temperatures so low that conventional metabolic rates would be glacial by human standards. Conventional instruments searching for carbon-based, water-reliant life might find nothing—not because the moon is sterile, but because we lacked the imagination to anticipate its biochemistry. Designing missions that could detect such alien life requires more than engineering expertise; it requires speculative models informed by chemistry, planetary science, and an openness to the unfamiliar.

Even the most advanced artificial intelligence systems, when applied to exoplanet data, are constrained by the biases embedded in their training. Algorithms trained on Earth-centric datasets will reproduce Earth-centric expectations. Without imaginative input—without human minds willing to propose the unorthodox—AI becomes a sophisticated mirror, reflecting only what we already know. To break new ground, the scientific method must be paired with the creative courage to hypothesise without immediate validation.

Some of the most thought-provoking developments in this area come from the intersection of science and speculative fiction. Visionary authors such as Stanisław Lem, Ursula K. Le Guin, and Liu Cixin have long explored the boundaries of biology, physics, and consciousness in ways that challenge scientific orthodoxy. Their imagined worlds, while fictional, often pose deeply rigorous questions: What defines intelligence? Could non-carbon life evolve sentience? Would alien perception alter not just how life is lived, but what reality is? These are not distractions from science—they are provocations that enrich it.

Philosophers of science have long recognised the legitimacy of such speculative reasoning. Karl Popper argued that bold conjectures, when subjected to falsification, are essential for scientific progress. Thomas Kuhn described scientific revolutions as shifts in paradigms—mental models that govern what questions are askable and what answers are admissible. Imagination, then, is not a soft alternative to scientific thought, but one of its deepest drivers. Without it, we might refine existing knowledge, but we will not fundamentally expand it.

At the practical level, training the next generation of scientists to value imagination alongside technical skill is crucial. Curricula that treat creativity as peripheral—confined to the arts or dismissed as unscientific—miss a vital dimension of discovery. Encouraging interdisciplinary collaboration, speculative modelling, and open-ended inquiry nurtures the mental flexibility required to confront phenomena that defy categorisation. After all, the cosmos is not obligated to conform to our current paradigms. If anything, the weight of evidence suggests it will continually surprise us.

To be clear, imagination must be tethered to logic, constrained by plausibility, and subject to empirical testing. It is not a licence for whimsy, but a discipline of envisioning the possible before it is observable. The point is not to fantasise without limits, but to expand the perimeter of the thinkable. As Arthur C. Clarke famously observed, "The only way to discover the limits of the possible is to venture a little way past them into the impossible." Science, at its best, does exactly that.

The search for alien life—be it through telescopes, rovers, or spectrometers—is ultimately a test of imagination. Not only must we identify potential habitats and plausible chemistries, but we must also confront our own cognitive constraints. Are we prepared to recognise life that thinks, metabolises, or evolves differently? Will we trust the strange when it doesn't resemble us? The answers to these questions will determine not just what we find, but how we interpret it.

Imagination is not an optional luxury in the scientific enterprise—it is a fundamental engine of insight. It allows us to conceive of quantum entanglement before it can be measured, to predict black holes before they can be photographed, and to model alien life before we know it exists. In a universe of infinite variability, clinging to the known is not caution—it is blindness. To move forward, we must learn to imagine responsibly, rigorously, and relentlessly. Only then will we be ready to meet whatever, or whoever, awaits us in the vast unknown.

Chapter 2
The Exoplanet Renaissance: How We Found Thousands of Worlds

For centuries, the stars were understood as distant suns, fixed in position and surrounded by mystery. Philosophers speculated, poets wondered, and scientists questioned—but evidence of other worlds remained elusive. It was only in the final decade of the twentieth century that the first exoplanet orbiting a main-sequence star was confirmed: 51 Pegasi b, a gas giant in a scorchingly close orbit. That detection, made in 1995 by Michel Mayor and Didier Queloz, was the whisper of a revolution. It cracked open the door to a universe more dynamic and populated than we had dared to imagine. What followed was nothing less than a renaissance—an explosion of discovery that redefined the cosmos not as a void between stars, but as a galaxy of planets. Thousands of them. And counting.

At the heart of this revolution was a spacecraft no larger than a small car: NASA's Kepler Space Telescope. Launched in 2009, Kepler was designed with a singular mission—to stare at a fixed patch of sky and observe the subtle dimming of stars as planets passed in front of them. This method, known as the transit technique, transformed planetary science. In its original four-year survey, Kepler observed over 150,000 stars and identified more than 2,600 confirmed exoplanets, with thousands more candidates awaiting verification. These were not speculative hints or mathematical maybes—they were solid, measurable worlds, many of them smaller than Neptune and orbiting sun-like stars. Kepler provided not just discoveries, but statistical clarity: planets are not rare, they are ubiquitous. Virtually every star, it turns out, hosts a planetary system of some kind.

But Kepler was more than a machine—it was a methodological leap forward. It exemplified a new era of astronomy in which the detection of planets no longer relied on chance alignments or indirect inference. Instead, it utilised high-precision photometry to measure the minuscule dip in starlight—often less than one percent—that occurs when a planet transits its host star. By observing multiple transits, astronomers could determine a planet's orbital period, size, and even hints of its atmosphere. When combined with radial velocity measurements—the so-called wobble method, which tracks a star's motion caused by gravitational tugs from orbiting planets—scientists could also estimate planetary masses. Together, these tools began to sketch not just dots in a catalogue, but detailed portraits of alien worlds.

The transit and radial velocity methods form the backbone of exoplanet detection, and they are both ingenious in their simplicity. The wobble method relies on the Doppler effect: as a star moves slightly towards and away from us, its light shifts to bluer or redder wavelengths. This allows astronomers to infer the presence of a companion exerting gravitational force. The transit method, on the other hand, is a study in shadowplay—watching the sky for the telltale blink of planetary eclipses. Both methods are limited by geometry: the transit technique works only when a planet's orbit aligns with our line of sight, while the radial velocity method favours larger planets with tighter orbits. Yet even with these limitations, the results have been staggering.

Among the most intriguing outcomes of the Kepler era was the identification of planetary types that defy our existing classifications. Chief among these are the so-called super-Earths and mini-Neptunes—planets larger than Earth but smaller than Neptune, with no analogues in our solar system. Some possess rocky surfaces and thick atmospheres; others may be gas dwarfs with substantial volatile envelopes. These intermediate planets blur the lines between terrestrial and gaseous worlds and challenge traditional models of planetary formation. Why are they so common elsewhere, but absent in our solar system? What processes shape their composition and evolution? These questions are not academic—they strike at the very core of how we understand planet-building.

In addition to expanding the typology of planets, the renaissance in exoplanet science has initiated a new frontier: the study of alien atmospheres. Using a technique called transmission spectroscopy, astronomers examine the light that passes through a planet's atmosphere during a transit. Different molecules absorb light at different wavelengths, creating spectral fingerprints that can reveal the presence of water vapour, methane, carbon dioxide, or other gases. This method is delicate, requiring immense precision and careful modelling, but it holds extraordinary promise. Already, transmission spectra have hinted at clouds of exotic composition, hazes thick with hydrocarbons, and atmospheres distorted by extreme radiation.

Perhaps the most tantalising application of this technique is the search for biosignatures—chemical signs that suggest biological activity. The detection of oxygen, ozone, or methane in specific combinations could indicate a living biosphere, though caution is warranted. Many such gases can also be produced abiotically, and interpreting spectral data involves numerous assumptions about atmospheric dynamics, stellar radiation, and planetary chemistry. Still, the fact that we can now read the skies of other worlds—deciphering their compositions, temperatures, and weather systems—is itself a scientific triumph. It shifts the conversation from the existence of planets to the character of planets, from abstract points to places with meteorology and potential habitability.

Yet for all we have discovered, we remain humbled by what we do not yet understand. Thousands of confirmed planets and tens of thousands of candidates have given us data—but not always clarity. The diversity of planetary systems is bewildering. We have found hot Jupiters skimming their stars in days, systems with multiple planets in tightly packed orbits, and others with enormous separations. There are planets that appear to have migrated inward after formation, reshaping their systems in violent upheavals, and others that seem to float in isolation, untethered to any star. Each discovery adds nuance but also complexity, reminding us that our solar system may be more unusual than once believed.

There are still glaring gaps in our knowledge. We struggle to detect small, Earth-sized planets in wide orbits—precisely the kind that might harbour life as we know it. Our sensitivity to atmospheric signatures is improving, but remains limited by both instrumentation and interpretation. The exoplanet census is impressive, but far from complete. The more we learn, the more the universe reveals itself to be not a mirror, but a mosaic—one in which Earth is merely one tile among billions.

The exoplanet renaissance is not a completed chapter, but an unfolding narrative. With the advent of next-generation telescopes like the James Webb Space Telescope, the European Extremely Large Telescope, and the proposed LUVOIR and HabEx missions, we stand at the edge of a new era—one in which exoplanet characterisation may become as detailed and varied as planetary detection once was. These instruments promise to peer into the atmospheres of Earth-sized worlds, to search for water and organic molecules, and perhaps even to capture direct images of distant planets. They will also test the limits of our imagination and demand greater integration of astrophysics, planetary science, and biology.

This chapter offers a guided tour through the methods and milestones that have defined the exoplanet renaissance. It examines how we've moved from detecting distant shadows to studying alien climates and assessing planetary diversity. It also urges caution and humility, recognising that every answer opens new questions, and every new technique uncovers new complexity. In many ways, this field is still in its infancy. But what a beginning it has been.

The stars no longer shine in isolation. They are accompanied by worlds—myriad, marvellous, and mysterious. Each discovery rewrites our place in the universe, turning what was once a philosophical speculation into an empirical fact. Planets are everywhere. The challenge now is to understand them—and to prepare ourselves for the possibility that among them, perhaps, one may harbour life.

Kepler's Legacy and the Rise of Planet Hunting

Before the early 1990s, the notion of planets orbiting distant stars was a matter of philosophical conjecture and speculative fiction rather than scientific consensus. Theories abounded, and indirect hints were debated, but no exoplanet had been unequivocally identified. Ours was a solar system of one, with planetary formation understood largely through the filter of local example. Then, in 1995, a seismic shift occurred: Michel Mayor and Didier Queloz announced the discovery of 51 Pegasi b, a gas giant orbiting a sun-like star. It was an astonishing find—not only because of the planet's existence, but because of its strangeness. Unlike anything in our own system, 51 Pegasi b orbited perilously close to its star, defying established models of planetary dynamics. The discovery marked the dawn of a new scientific era. Yet it would take the launch of a space telescope in 2009 to transform this dawn into a full-fledged renaissance.

The Kepler Space Telescope was not built to find one planet. It was built to find thousands. Named after Johannes Kepler, whose 17th-century laws of planetary motion revolutionised astronomy, the telescope carried a similarly transformative ambition: to statistically determine how common planets are in the Milky Way. Unlike previous instruments, Kepler was not aimed at individual stars. It was designed to monitor over 150,000 stars simultaneously, watching for the faint, periodic dips in brightness that would indicate a planet passing—or transiting—in front of its host star.

Kepler's observational method was deceptively simple but scientifically profound. A planet transiting its star blocks a small fraction of the starlight, typically less than one percent, resulting in a characteristic light curve. By measuring the depth, duration, and periodicity of these dips, astronomers could infer a planet's size, orbital period, and distance from the star. If a planet transited more than once, and the signal was consistent, it became a candidate for further analysis. And if the signal passed additional tests—including elimination of false positives from binary stars or instrumental noise—it could be confirmed as a bona fide exoplanet.

This approach represented a conceptual shift in planet hunting. It was no longer about lucky one-off detections, but about large-scale surveys and statistical power. Kepler allowed astronomers to move from anecdote to demographic insight. The results were staggering. Over its initial four-year mission, Kepler confirmed more than 2,600 exoplanets and identified thousands more candidates. Its discoveries included a wide variety of planetary types: Earth-sized worlds in habitable zones, compact systems with multiple planets in tight orbits, and planets so unlike anything in our solar system that they defied classification.

Perhaps the most profound impact of Kepler was not the individual planets it discovered, but what those discoveries implied. Kepler's data enabled scientists to estimate that there are more planets than stars in our galaxy—a realisation that fundamentally alters our place in the cosmos. It suggests that Earth is not an outlier but one of many possible configurations. The Milky Way, long thought to be a silent sea of stellar isolation, now appeared to be teeming with planetary diversity.

Kepler also played a pivotal role in revealing just how atypical our solar system might be. Many of its discoveries included planets known as super-Earths and mini-Neptunes—worlds with sizes between Earth and Neptune, yet largely absent from our own planetary neighbourhood. These planets appear to be the most common in the galaxy, and their prevalence challenges traditional models of planetary formation. In our solar system, a clear size gap exists between rocky terrestrial planets and gas giants. Kepler showed us that nature does not share this dichotomy. Instead, planetary systems can be tightly packed, with multiple planets orbiting within a few stellar radii of their star, often with surprising regularity.

The mission also yielded insights into planetary system architecture. Systems like Kepler-11 and Kepler-90 demonstrated configurations markedly different from ours—compact, dynamically stable, and often containing several planets in orbital resonances. Kepler-90, for example, boasts eight confirmed planets, tying with our solar system for the most planets around a single star, yet its layout is more akin to that of Jupiter's moons than to the spread of planets in our own system. These findings suggest that solar system formation is not bound to a singular blueprint, but is shaped by a multitude of processes and conditions that can produce a rich diversity of outcomes.

Crucially, Kepler didn't work alone. Its discoveries were strengthened and expanded through ground-based follow-up observations and other space missions. Radial velocity measurements provided mass estimates, allowing researchers to determine planetary densities and infer compositions. Some planets, like Kepler-10b, were found to be rocky and dense, while others, such as Kepler-22b, appeared to have thick atmospheres and potentially liquid water. The coordination between observational strategies marked a maturation in the field—planet hunting was no longer experimental; it had become systematic.

Beyond pure science, Kepler's legacy lies in its cultural and philosophical impact. For the first time, we had data suggesting that Earth-like planets in habitable zones are not merely possible—they are likely. The discovery of Kepler-186f, an Earth-sized planet in the habitable zone of its star, resonated deeply with both scientists and the public. It offered a tangible foothold for age-old questions: Are we alone? Is life a cosmic fluke, or a common consequence of planetary physics?

Yet Kepler's mission was not without limitations. It was confined to a fixed patch of sky, about 115 square degrees in the constellations Cygnus and Lyra. This meant that while its dataset was deep, it was not broad. Moreover, the transit method favours short-period planets—those that orbit close to their stars and transit frequently. Long-period, Earth-analogues are harder to detect, requiring longer observation times and more precise instruments. Despite these constraints, Kepler laid the groundwork for what followed. Missions such as TESS (Transiting Exoplanet Survey Satellite) and CHEOPS (Characterising Exoplanets Satellite) build directly on Kepler's foundation, expanding the search across the whole sky and refining the characterisation of known planets.

Perhaps the most poetic dimension of Kepler's legacy is its embrace of statistical uncertainty. While each individual detection is meticulously analysed, the mission's real strength lies in its capacity to quantify the unknown. By measuring occurrence rates, false positive probabilities, and transit depth distributions, Kepler enabled a new kind of astronomy—one that trades certainty for probabilistic insight, and that acknowledges ambiguity as a necessary part of expanding knowledge.

As Kepler's fuel ran out in 2018, the mission was formally declared over. But its influence endures, not only in the raw numbers—thousands of confirmed planets—but in the intellectual shifts it catalysed. It reframed the cosmos not as an empty expanse punctuated by stars, but as a dynamic theatre of planetary activity. It challenged long-held assumptions, inspired a generation of scientists, and opened new avenues of exploration that stretch far beyond our solar system.

Kepler's legacy is not simply a list of discoveries. It is a demonstration of what is possible when instruments, ideas, and imagination converge. It showed us that planets are common, that planetary systems are diverse, and that the conditions for habitability may not be rare. It taught us how to look, what to look for, and how to interpret the faintest whispers of distant worlds.

The rise of planet hunting, ushered in by Kepler, is still in its early chapters. The story continues through new missions, new technologies, and new frameworks of thought. But it was Kepler that dared to ask a question of scale: not whether planets exist, but how many, and what kinds, and what they might mean for us. That question, once rhetorical, is now empirical—and its answers are still unfolding, one light curve at a time.

Transit, Wobble, and Light Curves: How We Know What We Know

In the great cosmic detective story of our time—uncovering planets around distant stars—astronomers have become investigators of light and gravity, piecing together invisible worlds from subtle clues scattered across the sky. These planets do not reveal themselves in photographs; they cannot be seen directly through a telescope like the rings of Saturn or the craters of the Moon. Instead, their presence is inferred through delicate signatures embedded in starlight—an infinitesimal dimming, a rhythmic motion, a shift in spectral colour. Like footprints in the dust or a door left ajar, these telltale signs provide glimpses of the unseen. This is the art and science of exoplanet detection, and it is built upon the core techniques known as transit photometry, radial velocity, and light curve analysis.

To grasp the logic behind these methods, imagine the astronomer not as a passive observer, but as a patient detective who studies patterns and anomalies to deduce the existence of hidden agents. Every star is a suspect. Each flicker of brightness, every spectral twitch, is a clue. What began as speculation has now become a disciplined craft, with methods precise enough to detect planets smaller than Earth orbiting stars hundreds of light-years away.

Among these techniques, the transit method stands out for its elegance and effectiveness. It relies on a simple premise: if a planet crosses in front of its host star, as seen from our vantage point, it will block a small fraction of the star's light, producing a slight but measurable dimming. This event is called a transit, and when it repeats at regular intervals, it signals the presence of a planet in orbit. The amount of light blocked tells us the planet's size, while the time between transits reveals its orbital period. With additional data, scientists can even estimate the planet's distance from its star and its likely surface temperature.

This method revolutionised exoplanet science. The Kepler Space Telescope, launched in 2009, employed transit photometry to monitor more than 150,000 stars continuously, looking for these periodic dips in brightness. Over the course of its mission, Kepler identified thousands of exoplanets, many in multi-planet systems, offering unprecedented insight into the diversity of planetary architectures. Crucially, transit photometry allowed scientists to detect smaller, Earth-sized planets—objects too subtle for earlier techniques. It marked the first time that statistics, not serendipity, drove planetary discovery.

Yet even the most refined transit light curves require interpretation. Not every dimming event is caused by a planet. Stellar variability, binary star systems, and instrumental noise can all mimic planetary signals. To separate the genuine from the spurious, astronomers perform follow-up observations and use algorithms to model the light curves, examining their symmetry, depth, and duration. A true planetary transit produces a distinctive U-shaped curve: a steady dip followed by a return to baseline brightness. The precision of this signal has made the transit method the cornerstone of modern exoplanet surveys.

However, transits only reveal part of the picture. They tell us how big a planet is, but not how heavy. To calculate a planet's mass, astronomers turn to the radial velocity method, often referred to as the "wobble method." This technique is based on the gravitational interplay between a star and its planets. As a planet orbits, it exerts a gravitational pull on its star, causing the star to move in a small orbit of its own. This motion induces shifts in the star's spectral lines due to the Doppler effect—when the star moves towards us, its light is slightly blueshifted; when it moves away, it is redshifted.

By measuring these periodic shifts in stellar spectra, scientists can infer the presence of a planet and estimate its minimum mass. Unlike the transit method, which depends on a fortunate alignment, the radial velocity technique can detect planets regardless of orbital orientation, provided the motion produces a line-of-sight velocity change. It was this method that revealed 51 Pegasi b—the first exoplanet discovered around a main-sequence star—ushering in the modern age of exoplanet science.

When transit and radial velocity data are combined, they provide a fuller picture. Size from transits and mass from wobbles allow scientists to calculate planetary density—an essential clue to composition. A high-density planet may be rocky like Earth; a lower-density world is likely to be gaseous. Through this synergy, astronomers can begin to classify exoplanets not only by their orbits, but by their internal structures: super-Earths, gas dwarfs, hot Jupiters, and beyond.

More recently, a powerful refinement has emerged: the analysis of light curves for signs of planetary atmospheres. During a transit, a fraction of the star's light passes through the thin rim of the planet's atmosphere. Molecules in that atmosphere absorb light at specific wavelengths, imprinting spectral fingerprints onto the starlight. By comparing the star's spectrum during and outside of transit, astronomers can extract information about atmospheric composition. This method—known as transmission spectroscopy—has revealed sodium in the skies of hot Jupiters, water vapour on temperate planets, and complex hazes that obscure deeper analysis. Although still in its infancy, this technique promises to be one of the most powerful tools in the search for biosignatures.

Such atmospheric detections require extraordinary sensitivity. The variations in starlight can be as subtle as one part in ten thousand, akin to detecting the flicker of a firefly in front of a floodlight. Instruments like the Hubble Space Telescope and, more recently, the James Webb Space Telescope have begun to meet this challenge. Webb's infrared capabilities are particularly well suited to probing planetary atmospheres, especially for molecules like carbon dioxide, methane, and water—substances associated with life as we know it.

Yet, even as our tools improve, the detective work remains complex. Light curves can be deceptive. Stellar spots, pulsations, and magnetic activity can mimic or mask planetary signatures. Some false positives are nearly indistinguishable from real planets without extensive follow-up. The community has responded with increasingly sophisticated modelling, machine learning algorithms, and a growing network of ground-based observatories to validate space-based detections.

Beyond transits and radial velocities, other methods are emerging. Direct imaging, though extremely difficult, allows us to capture actual pictures of exoplanets by masking out the star's light. Gravitational microlensing uses the warping of spacetime to detect planets around distant stars when a foreground star momentarily magnifies the light of a background one. Each method has its own strengths and weaknesses, its own context in which it shines. But together, they form a toolkit—a set of complementary strategies that allow astronomers to read the sky like a forensic archive.

The detective metaphor is more than a rhetorical device. It captures the essence of scientific reasoning—inferring the presence of what cannot be directly seen, building models from partial evidence, and testing hypotheses against shifting data. It also reminds us that our knowledge is provisional, subject to revision as new techniques and technologies emerge. We are piecing together a vast cosmic mosaic from fragments of light, spectral distortions, and gravitational nudges. Each discovery is not an endpoint, but a lead to be followed.

Understanding these methods is not just about appreciating scientific ingenuity; it's about recognising how knowledge is built—slowly, collaboratively, and with reverence for uncertainty. Planet hunting is a triumph of both precision and imagination. It requires instruments that can measure movements smaller than a human hair from light-years away, and minds willing to interpret those movements as the silent choreography of worlds.

What makes this pursuit extraordinary is not just its technical complexity, but its philosophical depth. Through transits, wobbles, and light curves, we have learned that planets are not rare; they are the rule. Every flicker in a light curve might be a world. Every spectral shift might hint at weather systems, oceans, perhaps even life. These methods, so quietly methodical, are in fact seismic in their implications. They have turned our species from planetary residents into galactic cartographers, scanning the stars for signs of kinship in the dark.

And in doing so, they remind us that knowledge—true knowledge—emerges not from the obvious, but from the overlooked. From the shadow of a passing planet. From the spectral heartbeat of a distant sun. From the quiet confidence that, with the right tools and questions, we can decipher the invisible.

The Super-Earths and Mini-Neptunes Mystery

For centuries, our understanding of planetary formation was shaped by the familiar architecture of our solar system. It offered a clear hierarchy: small, rocky planets close to the Sun; gas giants in the outer regions; and icy bodies populating the distant margins. This configuration provided the foundational framework for planetary science, with the assumption that other planetary systems—should they exist—would follow a similar script. That assumption has now been thoroughly dismantled. The discovery of a class of planets known as super-Earths and mini-Neptunes—worlds that fall between Earth and Neptune in size, yet are absent from our own system—has not only expanded the planetary catalogue but disrupted the very principles upon which our models were built.

These intermediate-sized planets, ranging from about 1.5 to 3.5 Earth radii, occupy a curious no-man's-land in the traditional classification of planets. They are too large to be considered terrestrial like Earth or Mars, yet too small to be categorised with gas giants such as Neptune or Uranus. Until the advent of modern exoplanet surveys, especially the Kepler mission, there was no observational precedent for them. Yet they have turned out to be among the most common types of planets in the galaxy. This revelation has forced astronomers to ask an uncomfortable question: why does our solar system lack its most prevalent planetary neighbour?

Kepler's data revealed that roughly thirty to fifty percent of Sun-like stars host at least one planet in the super-Earth or mini-Neptune category. Some of these worlds orbit close to their stars, completing a full revolution in a matter of days. Others exist in multiple-planet systems packed into compact configurations, often with orbital resonances that provide clues to their dynamical evolution. This level of ubiquity suggests that super-Earths and mini-Neptunes are not exceptions—they are the statistical norm. And that realisation has profound implications for both planet formation theory and the search for habitable environments.

One of the first questions raised by these findings is whether super-Earths and mini-Neptunes are distinct categories or part of a continuum. Initially, the terms were used interchangeably, referring broadly to planets larger than Earth but smaller than Neptune. However, more refined data—particularly from transit and radial velocity measurements—has revealed a bifurcation in the population. Around 1.5 Earth radii, there appears to be a divide in planetary composition: below this threshold, planets tend to be rocky; above it, they often possess substantial atmospheres of hydrogen and helium. This gap, sometimes called the "radius valley," suggests divergent formation and evolutionary histories, potentially driven by processes such as photoevaporation, core accretion, or atmospheric loss due to stellar irradiation.

The distinction is not merely semantic. A super-Earth may be a massive rocky planet with a thin atmosphere, possibly featuring continents, tectonics, and conditions suitable for life. A mini-Neptune, by contrast, may have a deep envelope of gas and volatile materials, high internal pressures, and a surface—if it exists at all—buried beneath crushing atmospheric layers. Determining the internal structure of these planets requires precise measurements of both mass and radius, from which density can be inferred. Yet even with this information, uncertainty remains. Similar densities can result from different combinations of rock, metal, water, and gas. As a result, scientists must rely on models that simulate various configurations and test them against observational data.

One notable example is the exoplanet GJ 1214b, discovered in 2009. Roughly 2.7 times Earth's radius, it orbits a red dwarf star just forty light-years away. Its low density suggests that it is not a purely rocky world, but possesses a thick atmosphere or a significant water layer—or both. Follow-up studies, including transmission spectroscopy, revealed a featureless spectrum, likely caused by high-altitude clouds or haze that obscure deeper atmospheric layers. Despite this ambiguity, GJ 1214b remains a touchstone in the study of mini-Neptunes: an archetype of a planetary class with no analogue in our own system, yet a common outcome of planetary evolution elsewhere.

Another compelling case is Kepler-138, a system with multiple planets straddling the transition between rocky and gaseous compositions. Recent observations with the Hubble and Spitzer space telescopes suggest that some of its planets have densities consistent with a significant water component—so-called "water worlds" with vast oceans hundreds of kilometres deep. These environments defy easy comparison with Earth, but they raise tantalising possibilities. Could such planets host life beneath their global seas? Could they sustain subsurface ecosystems akin to those hypothesised on Europa or Enceladus?

What is particularly striking about these intermediate planets is how they challenge long-held models of planetary formation. According to the classical core accretion theory, a planet forms by accreting solid material until it reaches a critical mass, at which point it begins to rapidly attract gas from the surrounding protoplanetary disc. This runaway gas accretion leads to the formation of gas giants. The theory predicts that planets should either remain small and rocky or grow quickly into giants, with little stable ground in between. Yet super-Earths and mini-Neptunes occupy that intermediate ground, suggesting that the gas accretion process may be more gradual or subject to additional factors such as timing, disc density, and migration.

Some researchers propose that these planets formed farther out in their systems and migrated inward. This would explain their tight orbits and, in some cases, unusual densities. Others suggest that photoevaporation—the stripping of a planet's atmosphere by intense stellar radiation—could account for the differences observed. A mini-Neptune, subjected to high-energy ultraviolet light from its host star, might gradually lose its gas envelope, becoming a super-Earth. This theory is supported by observations showing that planets closer to their stars tend to be smaller and denser, consistent with atmospheric loss.

Still, none of these explanations fully account for the diversity and distribution of super-Earths and mini-Neptunes. Some appear to retain thick atmospheres despite close proximity to their stars. Others, located farther out, seem to have lost much of their volatile content. This inconsistency points to a complex interplay of formation location, migration history, and stellar activity—an interplay that varies from system to system, defying any singular model.

The absence of super-Earths and mini-Neptunes in our solar system adds another layer of complexity. Why did our planetary system leap from Mars-sized terrestrials to Neptune-sized gas giants, bypassing the most common planetary category in the galaxy? Some have suggested that Jupiter's early migration may have played a disruptive role, scattering planetesimals and preventing the formation of intermediate-mass worlds. Others argue that the solar nebula's composition or longevity may have limited gas accretion in the inner system. Whatever the explanation, the uniqueness of our planetary configuration invites humility. We are not the norm. If anything, we are the outliers.

This realisation carries profound implications for the search for life. If the majority of planetary systems host super-Earths and mini-Neptunes, then these planets may be the most likely sites for extraterrestrial biology. Yet they also pose challenges. Their thick atmospheres, high pressures, and potential lack of surface landmasses complicate the prospects for habitability. Even so, they cannot be dismissed. Life, as we continue to learn, is adaptable—and the cosmos appears to be replete with environments that test the limits of that adaptability.

Our expanding knowledge of these intermediate planets also sharpens our observational strategies. The next generation of telescopes—both space-based and terrestrial—will target super-Earths and mini-Neptunes for detailed atmospheric characterisation. Missions like the James Webb Space Telescope, the European Space Agency's ARIEL mission, and the upcoming Extremely Large Telescope aim to probe their skies for molecular signatures, thermal profiles, and cloud structures. The goal is not just to catalogue these worlds, but to understand their climates, chemistries, and potential for hosting life.

Ultimately, the mystery of super-Earths and mini-Neptunes is not merely a technical puzzle—it is an invitation to reimagine the boundaries of planetary science. These worlds, so common elsewhere and so absent here, force us to confront the parochialism of our assumptions. They remind us that planetary systems are shaped by myriad forces—some familiar, others still poorly understood—and that our solar system, long considered a model, may be an exception rather than a rule.

The more we study these planetary hybrids, the clearer it becomes that they are not anomalies but archetypes. They are rewriting the narrative of planetary formation, reshaping the targets of astrobiological research, and challenging us to expand our conception of what a planet can be. In doing so, they are not only enriching our understanding of the cosmos but also reframing our understanding of ourselves—as one planetary system among countless others, each carrying its own evolutionary story.

Exoplanet Atmospheres and the Art of Reading Alien Skies

In our search for life beyond Earth, few frontiers are more promising—or more technically demanding—than the study of exoplanet atmospheres. It is one thing to detect a distant planet orbiting a star hundreds of light-years away; it is quite another to tease apart the chemical composition of its atmosphere from the faintest variations in starlight. Yet this is precisely what astronomers are learning to do, using a technique that blends the rigour of physics with the art of inference: spectroscopy. Through this method, we are not only cataloguing planets but attempting to characterise them—probing the gases that swirl in their skies, identifying signatures of climate and chemistry, and, most tantalisingly, searching for the elusive markers of biological activity.

The fundamental premise of spectroscopy is deceptively simple: every molecule absorbs and emits light in a unique pattern, like a spectral fingerprint. When light passes through or reflects off a substance, certain wavelengths are absorbed, leaving behind dark lines or bands in the spectrum. These absorption features can be analysed to identify the molecules present in the atmosphere, from water vapour to carbon dioxide, methane to ozone. The challenge lies in detecting these signatures amid the overwhelming glare of the host star, especially when the planet itself is far too faint and small to be resolved directly.

To overcome this challenge, astronomers rely on the technique of transmission spectroscopy. When a planet transits—or passes in front of—its host star, a small portion of the starlight filters through the thin shell of its atmosphere. This light, having interacted with atmospheric molecules, carries the imprint of their absorption features. By comparing the star's spectrum during transit to the spectrum when the planet is not in front of the star, scientists can isolate the contribution from the planet's atmosphere.

The depth and shape of the absorption features in this differential spectrum reveal which gases are present and in what quantities.

The precision required for this method is extraordinary. The atmospheric signal typically represents a change of just one part in ten thousand—or less—in the total starlight received. It is the astronomical equivalent of detecting the outline of a mosquito wing against a spotlight, using only its effect on the light passing through it. Yet with the aid of space-based observatories like the Hubble Space Telescope, and now the James Webb Space Telescope, such measurements have become feasible. Webb's infrared capabilities, in particular, are ideally suited to detecting molecular features such as water, carbon monoxide, methane, and even more complex organics that may be linked to life.

One of the earliest and most influential examples of transmission spectroscopy was the analysis of HD 209458 b, a so-called "hot Jupiter" orbiting a star 150 light-years away in the constellation Pegasus. In 2001, astronomers using the Hubble Space Telescope detected sodium in the planet's atmosphere—a proof of concept that demonstrated the viability of atmospheric characterisation for exoplanets. Subsequent studies found evidence of hydrogen, oxygen, and carbon, as well as indications of atmospheric escape—a phenomenon where stellar radiation drives atmospheric particles into space. Though HD 209458 b is far too hot and gaseous to be considered habitable, its study laid the groundwork for similar techniques to be applied to smaller, cooler, more Earth-like planets.

Another landmark case was GJ 1214 b, a mini-Neptune orbiting a red dwarf just forty light-years away. Observations of its transmission spectrum yielded an unexpected result: a flat, featureless line. This lack of spectral features was interpreted as evidence of a high-altitude haze or cloud deck that obscures the lower atmosphere. While disappointing to some, this result was scientifically illuminating. It showed that exoplanet atmospheres are diverse and that clouds and aerosols—just as on Earth—can complicate spectral analysis. It also highlighted the importance of wavelength coverage: different molecules absorb light at different parts of the spectrum, and a truly comprehensive characterisation requires observations across a broad range of wavelengths.

As the catalogue of characterised exoplanets has grown, so too has the diversity of their atmospheres. Some hot Jupiters exhibit temperature inversions—stratospheres heated by the absorption of stellar ultraviolet radiation—while others appear to have clouds of exotic composition, such as silicates or iron. In cooler planets, researchers have detected water vapour and hints of carbon-bearing molecules. The variety of these atmospheres, shaped by differences in mass, temperature, stellar irradiation, and chemical history, is a testament to the complexity of planetary evolution. Each atmosphere is a case study in physics, chemistry, and environmental dynamics, offering insight not just into individual planets, but into the broader principles that govern planetary systems.

The ultimate ambition of atmospheric spectroscopy, however, is the detection of biosignatures—chemical indicators that suggest the presence of life. On Earth, the biosphere has transformed the atmosphere in measurable ways. Photosynthesis has filled the atmosphere with oxygen, while life processes produce methane, nitrous oxide, and a host of other trace gases. These gases, especially when found in thermodynamic disequilibrium, are considered potential biosignatures. For example, the simultaneous presence of oxygen and methane in significant quantities would be difficult to explain without invoking a biological source, since the two gases react readily and would not coexist unless constantly replenished.

Detecting such a combination on an exoplanet would be a seismic moment in science—an indirect but compelling argument for life beyond Earth. Yet the interpretation of biosignatures is fraught with ambiguity. Many of the gases we associate with life can also be produced through abiotic processes, such as volcanic activity, photochemical reactions, or mineral surface chemistry. Conversely, alien life may produce unfamiliar gases that we do not currently associate with biology. The risk of both false positives and false negatives looms large. As a result, scientists have called for a "biosignature framework"—an interdisciplinary, probabilistic approach that weighs multiple lines of evidence, including atmospheric composition, planetary context, and stellar environment.

This framework draws on lessons from Earth's own history. For much of its existence, Earth's atmosphere lacked oxygen, despite the presence of microbial life. The Great Oxygenation Event, around 2.4 billion years ago, marked a profound shift in atmospheric chemistry, driven by cyanobacteria. Understanding such transitions helps scientists recognise that life's imprint on an atmosphere may be subtle, variable, and time-dependent. Moreover, the biosignature gases we seek may only accumulate under specific conditions—sufficient planetary mass to retain an atmosphere, moderate stellar activity to avoid stripping it away, and climatic stability to allow complex chemistry to unfold.

The context in which we analyse an atmosphere is therefore as important as the atmosphere itself. A rocky planet in the habitable zone of a quiet star, with a stable climate and a protective magnetic field, presents a more promising target than a similar-sized world orbiting a flare-prone red dwarf. Stellar activity can inject noise into spectroscopic data, mimic or mask spectral features, and drive photochemical reactions that confound interpretation. Understanding the host star's behaviour and spectral output is thus essential to interpreting planetary spectra reliably.

As our observational capabilities expand, so too does our ambition. The James Webb Space Telescope, with its unparalleled sensitivity and spectral range, has already begun characterising atmospheres of small, temperate exoplanets. Future missions—such as ESA's ARIEL, the Nancy Grace Roman Space Telescope, and proposed direct imaging observatories like LUVOIR or HabEx—aim to push the frontier further, targeting true Earth analogues and searching for atmospheric signatures that could reveal life's presence. These missions will not only collect more data but do so with greater spectral resolution, broader wavelength coverage, and the ability to detect fainter signals.

Yet it is important to temper this optimism with epistemic humility. The detection of a biosignature will not yield an immediate declaration of alien life. It will raise a scientific hypothesis, to be tested, debated, and examined through multiple lenses. Just as the field of astronomy has matured from counting planets to characterising them, so too must our thinking evolve from searching for "Earth 2.0" to appreciating the full spectrum of planetary diversity. Life, if it exists elsewhere, may leave traces in atmospheres we cannot yet decipher or in combinations we have yet to consider.

In the meantime, the art of reading alien skies continues to sharpen. It blends astrophysics with chemistry, data analysis with theory, and a deep respect for the interplay between light and matter. It demands patience, precision, and imagination. Above all, it reaffirms the transformative power of perspective: that by studying the light from distant worlds, we are not just probing the cosmos—we are redefining what it means to observe, to understand, and perhaps, one day, to recognise life beyond Earth.

To read an alien sky is to engage in one of the most refined acts of inference science has ever attempted. It is a dialogue with the invisible, a listening for echoes in the light. And in that light, we may one day see ourselves—not in reflection, but in relation, as one form of life among many, scattered across the star-spangled vastness.

What We've Really Learned (And What We Haven't)

The past three decades have seen one of the most profound shifts in our understanding of the universe: the recognition that planets are not rare anomalies but the rule. Since the first confirmed discovery of an exoplanet orbiting a sun-like star in 1995, the field has grown from speculative beginnings to a mature scientific discipline supported by robust datasets, refined detection methods, and increasingly sophisticated models. We have identified more than 5,000 confirmed exoplanets to date, and statistical extrapolations suggest the Milky Way alone may harbour hundreds of billions. This knowledge has irrevocably changed the way we view our solar system and our place within it.

And yet, for all we have learned, the true impact of these discoveries may lie in what they reveal about the limits of our understanding. The sheer diversity of planetary systems—some strikingly orderly, others chaotic—has challenged long-held assumptions about how planets form, evolve, and persist. What was once imagined to be a tidy and predictable process now appears riddled with exceptions, anomalies, and outliers. The universe, it turns out, is not beholden to our neat theoretical frameworks.

One of the clearest lessons from the exoplanet revolution is that planetary systems are astonishingly diverse in both architecture and composition. Before Kepler, the expectation—derived from the architecture of our own system—was that terrestrial planets would reside close to their stars, gas giants would form in cooler outer regions, and orbital configurations would follow a broadly harmonious model. What Kepler revealed instead were compact systems with multiple planets orbiting in close quarters, gas giants known as "hot Jupiters" completing orbits in days, and planetary types that do not exist in our own system at all—such as super-Earths and mini-Neptunes. This diversity has not only expanded our catalogue of planetary possibilities but forced theorists to revisit the foundational models of planetary formation.

We've also learned that planets can exist in places once deemed improbable. Rogue planets—free-floating worlds not bound to any star—have been detected through gravitational microlensing, suggesting that planet formation may be more tumultuous and violent than previously thought, with ejections a common outcome. Similarly, planets have been found around binary stars, pulsars, and red dwarfs, broadening our definition of what constitutes a potentially viable environment for planet formation. These discoveries have highlighted nature's ingenuity in assembling worlds under widely varying conditions.

In terms of composition, we now appreciate that planet classification is far more complex than the traditional divide between "rocky" and "gaseous." Planetary density measurements have revealed a spectrum of internal structures, from iron-rich terrestrial planets to low-density, puffed-up giants and water-worlds with vast global oceans. Spectroscopic analyses of exoplanet atmospheres, while still in their early stages, have uncovered clouds of silicates, hazes of unknown composition, and chemical profiles shaped by temperature extremes, atmospheric escape, and irradiation. Every new data point adds nuance, suggesting that there are few—if any—universal rules.

We have, moreover, begun to realise the critical role that host stars play in shaping planetary destinies. The chemistry, age, activity level, and radiation output of a star directly influence planet formation, atmospheric retention, and surface conditions. Red dwarfs, once overlooked, are now prime targets in the search for habitable planets due to their longevity and prevalence. But they also present challenges, including intense flaring that may strip planetary atmospheres or prevent biological development. These insights underscore the interconnectedness of stellar and planetary evolution, and the importance of viewing planets not in isolation, but as participants in dynamic star-planet systems.

Crucially, we have also learned to appreciate the role of statistical science in shaping astronomical understanding. No single planet is definitive. Rather, it is the aggregation of thousands of observations, refined through Bayesian analysis and probabilistic modelling, that allows us to make generalisations. The Kepler mission didn't just detect planets; it quantified planetary occurrence rates, demonstrating that Earth-sized planets in habitable zones are relatively common. These statistics don't answer whether any are truly Earth-like in terms of climate or biology, but they do provide a fertile foundation for focused investigation.

At the same time, our limitations remain stark. For all the planets we have detected, the overwhelming majority remain little more than data points—known by mass, radius, or orbital period, but not yet truly characterised. Our knowledge of exoplanet atmospheres is confined largely to large, close-in worlds with bloated envelopes that make them easier to study. Earth-sized planets in habitable zones remain elusive, both because of the subtlety of their signals and the current limits of our instruments. We have glimpsed water vapour and possible clouds on a handful of worlds, but the detection of unambiguous biosignatures remains out of reach—promised, perhaps, by future missions, but not yet delivered.

The difficulty in characterising small, temperate exoplanets also reflects the limitations of our detection techniques. The transit method, for example, requires fortuitous orbital alignments, and biases us toward short-period planets. The radial velocity method is more sensitive to massive companions. Direct imaging, though developing rapidly, remains largely constrained to young, bright gas giants. Each method contributes a different piece of the puzzle, but none offers a full picture. As a result, we are left with a tantalising but incomplete understanding of what these worlds are really like.

Even the data we do have invites multiple interpretations. A flat transmission spectrum may indicate a cloud-covered atmosphere—or no atmosphere at all. A spectral feature might signify water, methane, or something more exotic. Our models are limited by their inputs, and often rely on assumptions derived from Earth-based chemistry and climate. This introduces a profound epistemological challenge: our ability to interpret alien worlds is constrained by our terrestrial perspective. The same bias that helped us build successful search strategies may now be limiting our capacity to imagine alternatives.

And then there is the matter of life. Despite the centrality of this question to public fascination with exoplanets, the field of astrobiology remains speculative. We understand very little about how life begins, adapts, or endures in non-Earth-like conditions. The so-called biosignature gases we seek may have false positives and false negatives. They may not even be relevant to alien biochemistries. If life exists elsewhere, it may look nothing like our own, leaving signatures we do not yet know how to detect. Alternatively, life may be so rare, so contingent upon improbable factors, that even with billions of worlds, the odds remain long. These are questions not of instrumentation alone, but of deep theoretical significance.

Despite these uncertainties, the exoplanet revolution has yielded something even more valuable than data: perspective. It has decentralised Earth from the cosmic narrative, not by diminishing its uniqueness, but by revealing its context. We now understand that planetary formation is common, that planetary diversity is vast, and that the conditions for habitability may arise in surprising places. This insight reshapes our existential orientation. It invites wonder and redefines isolation—not as a certainty, but as a question awaiting better answers.

It also invites intellectual humility. We are, by cosmic standards, just beginning. Our instruments are primitive, our catalogues incomplete, our models provisional. The field of exoplanet science sits at a thrilling intersection of discovery and ignorance. Each confirmed world is a revelation; each ambiguity a call to refine our tools and broaden our thinking. This interplay of certainty and uncertainty is not a flaw—it is the heartbeat of science itself.

So what have we really learned? That planets are everywhere. That they are more diverse than we imagined. That our solar system is not a template, but one case among many. That habitability may be more about context than about copying Earth. And perhaps most importantly, that the act of searching is as transformative as the discovery itself. For every confirmed detection, there are ten new questions. For every theory, a dozen counterexamples. Yet in this complexity lies beauty—because it reminds us that the universe is not obliged to be simple. It is obliged only to be real.

And what have we not yet learned? Most of it. But if the last few decades have taught us anything, it is that the universe rewards curiosity—not always with answers, but with richer questions. In that spirit, we continue to look outward, not to confirm what we already believe, but to confront what we do not yet understand. And in doing so, we move a little closer—not to certainty, but to comprehension. Not to finding ourselves elsewhere, but to seeing ourselves more clearly here.

Chapter 3
Worlds Without Suns: Rogue Planets and Strange Homes for Life

In our solar system, the Sun is the heartbeat—an incandescent anchor around which all planets revolve and from which life on Earth draws its energy. For centuries, we have imagined that without a star, a planet would be nothing more than a frozen husk drifting silently through the void. Yet recent theoretical advances and observational hints are inviting us to reconsider this assumption. What if the universe is teeming not only with Earth-like worlds bathed in starlight but with orphaned planets—rogue worlds set adrift in interstellar space? These starless planets, often thought to be hostile to life, may in fact be strange sanctuaries, hosting environments where biology clings on, nourished not by sunlight, but by heat from within.

Rogue planets are bodies that do not orbit a star. They wander the galaxy freely, unbound by gravitational ties to any stellar system. Some may have been ejected during the turbulent early stages of planetary system formation—gravitational interactions flinging them outwards like cosmic debris. Others may have formed in isolation, their mass insufficient to ignite nuclear fusion and become stars themselves. For a long time, such worlds were relegated to the periphery of serious scientific consideration. Without a sun, what could possibly survive there? Yet this question presumes that life is inextricably tied to light—a belief increasingly challenged by our expanding knowledge of extreme life on Earth and the rich physics of planetary interiors.

Observational evidence for rogue planets is growing, albeit slowly, given the inherent difficulty in detecting objects that emit little to no light. Gravitational microlensing—a technique that exploits the warping of space-time by massive objects—has provided some of the most compelling data. In 2011, researchers using the MOA and OGLE collaborations identified what appeared to be a population of Jupiter-mass objects without host stars, suggesting that such planets may be as common as stars themselves. Infrared surveys, though limited in scope, have hinted at free-floating planetary-mass bodies in young star-forming regions. The full extent of this population remains unknown, but the numbers are likely vast. The galaxy, it seems, may be not only filled with stars and their companions, but also with dark wanderers—planets that have no day, no year, no sunrise.

At first glance, such environments appear utterly inhospitable. Devoid of stellar energy, their surfaces would rapidly cool, plunging to temperatures near absolute zero. Yet this image, while dramatic, is misleadingly simplistic. Planets are not inert balls of rock; they are dynamic systems with internal heat sources. On Earth, geothermal energy powers volcanoes, hydrothermal vents, and even drives plate tectonics. This heat originates from residual formation energy and the decay of radioactive isotopes in the mantle and core. Rogue planets, especially those with substantial mass and insulation, could retain internal warmth for billions of years. If they possess thick atmospheres of hydrogen or methane, these could trap heat via greenhouse effects, further extending the period during which subsurface environments remain thermally active.

The potential for subsurface habitability is one of the most intriguing implications of rogue planets. Even in the absence of solar radiation, geothermal gradients can maintain liquid water beneath an insulating crust. Analogues exist here on Earth. Europa, Enceladus, and possibly Ganymede—moons of Jupiter and Saturn—harbour subsurface oceans beneath icy shells, kept warm by tidal forces and radioactive decay. Hydrothermal vents on Earth's ocean floor host complex ecosystems powered not by photosynthesis, but by chemosynthesis—organisms deriving energy from chemical reactions with minerals leaching from the planet's interior. These ecosystems are not merely rudimentary; they include shrimp, crabs, and tube worms, thriving in utter darkness. If such systems can flourish at the bottom of Earth's oceans, they might also arise in the lightless oceans of rogue planets.

The key requirement is energy—some mechanism to drive metabolism, enable chemical gradients, and maintain stable conditions for organic chemistry. On Earth, sunlight is dominant, but it is not exclusive. Chemotrophic life forms demonstrate that where there is energy flow, life can find a foothold. On a rogue planet with an active mantle and volatile-rich composition, hydrothermal systems could support microbial biospheres for extended timescales. Moreover, thick ice layers could protect these biospheres from cosmic radiation, much as the ocean shields deep-sea life on Earth.

More speculative still is the possibility of bioluminescent life—organisms that generate their own light for communication, predation, or mating. On Earth, bioluminescence is common in deep-sea creatures where sunlight does not penetrate. In the pitch-black oceans of a rogue world, such traits might become universal. Evolution in eternal night could yield a rich tapestry of sensory adaptations: infrared detection, chemical signalling, perhaps even sonar-like echolocation. These creatures would not only adapt to darkness—they would be shaped by it, embodying forms of intelligence and perception radically different from our own.

The role of gravitational heating and tidal flexing also deserves attention. While typically associated with moons orbiting gas giants, tidal forces can arise in multiple-body systems, such as a rogue planet with a large moon. The flexing of planetary interiors due to gravitational interactions can generate significant internal heat—enough to sustain geological and potentially biological activity. This is the mechanism thought to keep Europa's ocean liquid and Enceladus's geysers active. A similar dynamic on a rogue planet could extend its habitability zone far beyond the usual constraints of distance from a star.

What emerges from these models is a vision of habitability decoupled from stellar light. It is a profound paradigm shift. Rather than searching only for planets in the "habitable zone" around stars, we must consider a broader category of "ecospheres"—regions within planets themselves where conditions allow for stable, energy-supported chemistry. This perspective expands the potential locations for life dramatically and aligns with a more systems-based view of habitability—one that looks inward as much as outward.

There is also a psychological dimension to rogue planets that should not be ignored. These worlds, shrouded in darkness and wandering the interstellar void, evoke feelings of solitude and abandonment. They are metaphors as much as they are scientific entities—symbols of isolation, resilience, and the tenacity of life in adversity. In speculative fiction, rogue planets often serve as backdrops for narratives of survival, loss, or renewal. Scientifically, they challenge our heliocentric assumptions and invite us to consider forms of life—and modes of existence—that are not centred on warmth and light, but on heat, chemistry, and depth.

This confrontation with lifeless starscapes—worlds that orbit nothing, that rise and set to no dawn—forces us to ask deeper questions. What do we mean by "life-supporting"? How broad should our definitions be? And are we ready to recognise life that does not resemble our own? These questions are not merely academic. They guide the design of instruments, the framing of search strategies, and the interpretive models we bring to ambiguous data. They also shape our philosophical orientation—our readiness to accept that life may be common, but light-dependent life may not be.

As we continue to refine our observational capabilities, rogue planets may emerge not just as curiosities, but as laboratories for the study of life in extreme conditions. Missions that once sought Earth-like planets around Sun-like stars are now beginning to contemplate the darker corners of the cosmos. In the years ahead, gravitational microlensing surveys, infrared observatories, and perhaps even interstellar probes may reveal more about these cold, lonely worlds—and about the astonishing resilience of life itself.

To contemplate rogue planets is to stretch the imagination, to discard the comforting centrality of the Sun and entertain the possibility that life can flourish in the unlikeliest of places. These worlds whisper a quiet truth: that warmth can come from within, that isolation does not preclude vitality, and that even in the void, there may be company. The cosmos, in its vast and varied architecture, seems to favour not uniformity but possibility. And in that possibility, rogue planets stand not as aberrations, but as reminders of how much we have yet to discover—and how far life might extend beyond the boundary of light.

The Nomads of the Galaxy

In the grand theatre of the cosmos, planets have long been imagined as celestial companions to stars—dancing through well-ordered orbits, tethered by gravity, and nourished by light. Our solar system has deeply informed this view, reinforcing the notion that starlight is a prerequisite not only for planetary stability but also for life. Yet the discovery and study of rogue planets—worlds that drift untethered through interstellar space—have profoundly unsettled this assumption. These nomads, wandering in eternal night, seem at first glance to be inhospitable wastelands. But growing evidence suggests that their apparent desolation may be deceptive. Some may carry within them the silent warmth of internal energy—geothermal heat capable of sustaining subsurface oceans, stable environments, and perhaps even life.

Rogue planets, also called free-floating planets, are celestial bodies that do not orbit any star. They are not anchored in gravitational ballet; they drift through the galaxy as solitary travellers. The origins of these planets are varied. Some may have formed in isolation from collapsing gas clouds, never reaching the mass required to ignite nuclear fusion like stars. Others likely began life in traditional planetary systems, only to be violently ejected by gravitational perturbations—perhaps from close encounters with other massive planets or passing stars during the early, chaotic epochs of planetary system formation. Regardless of their genesis, their abundance appears to be significant. Observational surveys using gravitational microlensing techniques suggest that the number of rogue planets may rival, or even exceed, the number of stars in the Milky Way.

The microlensing technique, which relies on the gravitational bending of light from a background star as a foreground object passes between it and the observer, is currently one of the most effective means of detecting rogue planets. In 2011, a collaborative study between the MOA (Microlensing Observations in Astrophysics) and OGLE (Optical Gravitational Lensing Experiment) projects presented evidence for a population of Jupiter-sized objects unbound to any star. These objects, observed through brief magnifications of distant starlight, hinted at a vast, hidden population of planetary-mass bodies wandering the galaxy. More recently, the Nancy Grace Roman Space Telescope, scheduled for launch in the coming years, promises to expand this census dramatically, with the sensitivity to detect even Earth-sized rogue worlds.

At first glance, rogue planets would seem unlikely candidates for habitability. Without a nearby star, they receive no significant stellar radiation. Their surfaces, exposed to the cold vacuum of space, would rapidly freeze. However, this view neglects the role of internal heat—a planetary attribute independent of starlight. On Earth, geothermal energy accounts for a small fraction of the total surface energy budget, dwarfed by the energy received from the Sun. But in the absence of stellar input, geothermal processes become paramount. The decay of radioactive isotopes such as uranium-238, thorium-232, and potassium-40 within a planet's mantle and core can generate substantial heat. This heat, trapped beneath insulating layers of rock, ice, or gas, could maintain subsurface reservoirs of liquid water for billions of years.

The feasibility of life on such a world hinges on the persistence and distribution of this heat. Consider Earth's own deep biosphere—a vast microbial ecosystem that thrives kilometres beneath the surface, entirely divorced from sunlight. These extremophiles derive energy from the chemical interactions between water and rock, particularly in hydrothermal systems. The discovery of such life forms, notably near hydrothermal vents on the ocean floor, was a paradigm shift in biology. It demonstrated that sunlight is not the sole driver of life, and that chemosynthesis—life powered by chemical reactions rather than photosynthesis—is not only possible, but remarkably robust.

If Earth's deep biosphere can persist without sunlight, it is conceivable that similar systems could emerge on rogue planets. Indeed, models suggest that a rogue planet with a mass similar to

Earth's, insulated by a thick atmosphere—especially one composed of hydrogen—could retain enough internal heat to keep an ocean beneath an ice shell in a liquid state. Hydrogen, being a light but effective greenhouse gas, could trap heat efficiently, allowing for temperatures suitable for liquid water tens or even hundreds of kilometres below the surface. In this scenario, the surface would be a frozen wasteland, but beneath the ice, a stable, pressurised ocean might provide the conditions necessary for life.

A compelling analogue exists in our own solar system: Europa, one of Jupiter's moons. Despite being far from the Sun, Europa is believed to harbour a global subsurface ocean beneath a shell of ice. This ocean is kept warm not by solar radiation, but by tidal heating—gravitational interactions with Jupiter and neighbouring moons that flex Europa's interior, generating heat. While a rogue planet would lack tidal interactions with a star, similar heating could be generated if the planet retained a sizeable moon, or through purely radioactive means. In both cases, the key insight is that heat can be generated and sustained internally, decoupling habitability from stellar proximity.

Further, the chemical potential for life on such planets is substantial. Water, carbon, nitrogen, and other life-essential elements are common throughout the galaxy, found in comets, interstellar dust, and planetary atmospheres. On a rogue planet with a differentiated interior, hydrothermal vents could supply reduced minerals into the ocean, creating the redox gradients that power chemosynthetic metabolisms. Over time, these conditions could foster microbial ecosystems adapted to high pressure, low light, and chemical energy. The evolutionary pressures in such environments would likely produce life forms radically different from those on Earth—organisms shaped by darkness, chemistry, and slow metabolic rates.

The prospect of life on rogue planets also forces a reassessment of the traditional habitable zone—the so-called "Goldilocks zone" around a star where liquid water can exist on a planet's surface. This concept, while useful, is inherently heliocentric. It assumes that stars are the primary, perhaps exclusive, sources of planetary heat. Rogue planets challenge this notion, suggesting that habitability may be far more widespread, if we redefine it in terms of energy and stability rather than mere distance from a star. They broaden the search parameters for life and expand the conceptual canvas of astrobiology.

Scientifically, the study of rogue planets is still in its infancy. Their detection is difficult, and their characterisation even more so. Direct imaging is almost impossible due to their faintness, and transit methods—relying on a planet crossing in front of a star—are inherently inapplicable. Yet the progress made through microlensing, combined with future space-based observatories and potentially even interstellar probes, may eventually allow us to study these elusive worlds in greater detail.

Their philosophical significance, meanwhile, is immediate. Rogue planets are symbols of resilience in isolation—worlds that thrive, or at least endure, without the guiding presence of a star. They invite us to imagine life not as a product of ideal conditions, but as a tenacious process that exploits whatever energy gradients exist. In the cold, dark corners of the galaxy, these planets may carry within them the fire of life—a slow, hidden metabolism that endures against the odds.

In contemplating these nomads, we are reminded of a deeper truth: that life may not be limited by our expectations. The universe, after all, is not built to suit our imaginations. It is built on possibility. And among its most profound possibilities is the idea that life, in some form, may exist even where light cannot reach—thriving not in radiance, but in the quiet warmth of a planet's heart, drifting silently through the stars.

Underground Oceans and Subsurface Sanctuaries

When we think of alien worlds that could host life, it is tempting to imagine planets basking in gentle starlight, orbiting calmly within the habitable zones of distant suns. This model, drawn from the example of Earth, is tidy and reassuring. Yet the cosmos rarely adheres to our expectations. Recent advances in planetary science suggest that some of the most promising environments for life might be hidden, not in the warmth of solar radiation, but deep beneath layers of rock and ice—entire biospheres concealed in underground oceans, sheltered from the vacuum of space and sustained not by light, but by heat rising from within.

Nowhere is this idea more vividly realised than in the emerging models of subsurface oceans on rogue planets—those untethered wanderers that drift through interstellar space, unlit by any star. While their surface conditions would be inhospitably cold, potentially nearing absolute zero, their interiors may tell a different story. If these planets possess sufficient internal heat, generated through radioactive decay or residual formation energy, they could sustain liquid water oceans beneath insulating ice shells. In such hidden sanctuaries, shielded from cosmic radiation and isolated from the chaos of stellar activity, life might find a quiet foothold.

The plausibility of subsurface oceans on rogue planets draws heavily from what we have already learned about similar environments within our own solar system. Moons such as Europa, Enceladus, Ganymede, and possibly Titan have shown compelling evidence for global oceans beneath icy crusts. These moons orbit gas giants rather than stars, and their oceans are kept warm not by solar radiation, but by tidal forces—gravitational interactions that flex their interiors, generating frictional heat. Although rogue planets lack a parent star to create tidal effects on this scale, they could retain sufficient internal heat through radioactive decay and, if massive enough, might possess their own moons, creating localised tidal forces.

The physics behind this form of subsurface habitability is well understood. Water ice is an excellent insulator, and a thick shell of ice can preserve the heat beneath it for geologically significant timescales. Even a modest planet, roughly Earth-sized, with a mantle rich in heat-producing isotopes, could sustain a subsurface ocean for billions of years. If such a planet had formed with a thick atmosphere—particularly one rich in hydrogen—it could further retain surface heat through a greenhouse effect. But even in the absence of an atmosphere, the insulating capacity of ice alone might be enough to maintain a liquid ocean at depth.

This has led to the proposal of so-called "Stevenson worlds," named after planetary scientist David J. Stevenson, who first speculated in 1999 that Earth-sized rogue planets could maintain subsurface oceans under thick hydrogen atmospheres. Such worlds would be completely dark at the surface, but internally warm and potentially water-rich. Later models extended this idea, suggesting that even without an atmosphere, internal heat could maintain isolated aquatic environments—akin to subglacial lakes in Antarctica, such as Lake Vostok, where microbial life has been hypothesised to exist despite the total absence of sunlight.

The biochemistry of life in such settings would differ from the familiar photosynthesis-driven ecosystems of Earth's surface. Instead of deriving energy from sunlight, life forms in underground oceans would rely on chemosynthesis—extracting energy from the chemical reactions between water and rock. This is not merely speculative. On Earth, entire ecosystems flourish around deep-sea hydrothermal vents, where microbes oxidise hydrogen sulphide and other reduced compounds to produce organic matter. These communities support complex food webs, including tube worms, crustaceans, and molluscs, in environments completely devoid of light. Such systems demonstrate that life can thrive wherever there is a source of energy, liquid water, and chemical disequilibrium.

Transposing this model to a rogue planet's subsurface ocean raises several intriguing possibilities. Hydrothermal vents on the seafloor, fuelled by interactions between a silicate mantle and ocean water, could create hotspots of biological activity. These vents would offer heat, minerals, and chemical gradients—the building blocks of habitability. Over time, microbial colonies could evolve to exploit these niches, perhaps developing novel metabolic pathways adapted to low energy environments. Given the vast and relatively stable conditions of such oceans, evolution could proceed slowly but persistently, resulting in ecosystems with entirely alien biochemistries.

Some researchers have gone further, proposing that these environments might support not just microbial life, but complex organisms. The evolution of multicellularity, sensory perception, and even intelligence does not necessarily require sunlight; it requires selective pressure and sustained energy input. In a rogue planet's ocean, such pressures might come from predator-prey dynamics, competition for nutrients near vents, or environmental shifts caused by tectonic activity. While this remains speculative, it is no more improbable than the rise of life on Earth from a primordial soup.

From an observational standpoint, subsurface oceans on rogue planets are profoundly difficult to confirm. These worlds emit almost no light, and their biospheres, if they exist, are buried beneath kilometres of ice. Yet indirect detection methods are improving. Future infrared surveys may reveal the faint thermal signatures of internal heat escaping through ice. Gravitational microlensing surveys could provide population statistics, allowing us to estimate how common such worlds may be. The detection of chemical plumes, analogous to those observed on Enceladus, might one day be possible with next-generation space telescopes capable of ultra-sensitive spectroscopy. If rogue planets pass through interstellar clouds or interact with background stars, they may leave detectable chemical trails—signatures of active geology or even outgassing from subsurface oceans.

Beyond science, the concept of subsurface sanctuaries on rogue planets carries profound philosophical resonance. It suggests that life is not necessarily fragile or contingent on ideal conditions. Instead, life may be robust, opportunistic, and surprisingly adaptable—capable of taking root in the most concealed and unexpected places. It also redefines what we mean by "habitability." Rather than limiting our search to Earth-like conditions on the surface of temperate planets, we are beginning to embrace a broader, more inclusive view—one that values internal geodynamics, chemical potential, and the protective embrace of planetary ice.

This shift has practical consequences for astrobiology and planetary exploration. As we refine our models of planetary interiors, we may discover that underground oceans are not rare anomalies, but common planetary features. Subsurface water could exist not only on rogue planets, but also on moons, dwarf planets, and even large asteroids. The search for life, therefore, must expand downward as well as outward, into the hidden depths of planetary crusts and the isolated vaults of interstellar space.

In essence, the study of underground oceans challenges us to see beyond the obvious. It asks us to consider the invisible habitats that lie beneath icy exteriors, where silence and darkness reign but where life, in its quiet resilience, might flourish. These sanctuaries are not just geological curiosities—they are potential oases in a cosmos we once thought barren. As our technological and conceptual tools evolve, we may one day discover that the universe teems with life—not always basking in sunlight, but thriving in shadow, beneath the frozen skins of worlds that drift alone through the dark.

It is a vision at once humbling and inspiring. It reminds us that life's persistence is not necessarily tied to cosmic grandeur, but to the unassuming physics of pressure, heat, and chemistry. And it calls us to keep looking—not just at the stars above, but at the silent worlds that move among them, quietly harbouring secrets in their subterranean seas.

Bioluminescent Ecosystems: Evolution in Eternal Night

In the crushing darkness of Earth's deepest oceans, where sunlight cannot penetrate and temperatures hover near freezing, life thrives. At these depths—kilometres beneath the surface—light is not just absent; it is irrelevant. Yet within this realm, nature has evolved a remarkable alternative to visibility: bioluminescence. Through complex biochemical reactions, organisms produce their own light, illuminating the abyss with flashes, pulses, and ethereal glows. It is here, in these extreme environments, that we catch a glimpse of how life might adapt on planets far from any star. These glowing oceanic ecosystems may serve as analogues for what could evolve in subsurface oceans or the internal seas of rogue planets, where sunlight has never been and will never be.

The notion that life requires light is a deeply ingrained bias—an artefact of our surface-bound perspective. However, the vast majority of Earth's biosphere exists in darkness, from subterranean bacteria to deep-sea communities living in perpetual night. In these environments, energy is decoupled from photosynthesis. Chemosynthesis reigns supreme, fuelling life with energy derived from inorganic chemical reactions, such as the oxidation of hydrogen sulphide or methane. In the absence of visual cues, other adaptations emerge: heightened chemical sensitivity, pressure-resistant cell structures, and most intriguingly, the evolution of bioluminescence as both a sensory and communicative tool.

Bioluminescence, found across a wide array of marine species—from microscopic plankton to deep-sea squid and anglerfish—is a testament to evolution's ingenuity. It arises through the reaction of luciferin (a light-emitting molecule) with oxygen, catalysed by an enzyme called luciferase. The process is highly efficient, producing very little heat, and is often regulated by complex neural or biochemical pathways. Bioluminescence serves diverse functions: luring prey, attracting mates, deterring predators, and even facilitating camouflage through counter-illumination. In a world where light is a rare commodity, its emission becomes a powerful evolutionary currency.

Transposing this principle to alien ecosystems, particularly those on rogue planets or beneath thick ice crusts, invites both speculation and scientific insight. If such planets harbour subsurface oceans heated by geothermal or tidal energy, and if life arises in those oceans, it is plausible that some form of bioluminescence could evolve. In these eternally dark worlds, where visual systems might otherwise be obsolete, the emergence of self-generated light could represent a breakthrough in environmental awareness, communication, and predation.

Consider the case of a rogue planet—untethered from any star and enveloped in darkness. If it retains internal heat sufficient to support a liquid ocean beneath an icy shell, hydrothermal activity on its ocean floor could provide the energy necessary for life. Microbial mats might form near vents, chemosynthetic bacteria could flourish, and higher-order organisms could evolve around nutrient-rich hotspots. As life becomes more complex, the selective pressures for navigation, territory defence, mate location, and prey capture would intensify. In such an environment, the ability to emit light could be a decisive evolutionary advantage.

Parallels on Earth reinforce this scenario. The anglerfish, for example, uses a luminescent lure to attract unsuspecting prey. Certain species of shrimp and squid emit bright flashes to confuse predators or facilitate escape. Some deep-sea fish even possess light-producing organs arranged in patterns that mimic schools of smaller fish—creating a form of optical illusion. In the context of a dark alien ocean, similar evolutionary strategies could emerge. Life might evolve specialised organs to control wavelength, intensity, and rhythm of emitted light, adapting these signals for different functions—from silent courtship to strategic deception.

But beyond mere survival, bioluminescence on alien worlds could also shape the very nature of sensory perception. Earth-based organisms in the deep ocean often combine visual and chemical cues to interpret their environment. On a lightless planet, chemical gradients, temperature shifts, and sound may dominate early sensory evolution. Yet once bioluminescence evolves, it introduces a new channel of information—one that enables spatial awareness at a distance, individual recognition, and dynamic signalling. Over evolutionary time, this could lead to the development of sophisticated light-based languages, patterned displays for group coordination, or even rituals that mark the rhythm of life in a place where celestial cycles are meaningless.

The implications for intelligence are profound. Light is not only a medium for seeing; it is a medium for abstraction. On Earth, bioluminescence has evolved independently over forty times—a testament to its adaptive utility. If alien organisms evolved the ability to control light, they might begin to use it not merely for practical functions but for symbolic ones. Flash patterns could encode information, memory, or intent. In such a world, light could become a substrate for culture—pulses passed between individuals forming a kind of bioluminescent script, shaping collective behaviour, myth, and possibly even something resembling language.

Yet such speculations must be anchored in biology. The evolution of bioluminescence requires metabolic cost. It must offer sufficient benefit to be retained and refined. On Earth, this cost is offset by the advantages of survival and reproduction in highly competitive ecosystems. In alien oceans, the same principle would apply. Only if the energy return—be it through successful hunting, mating, or predator evasion—outweighs the metabolic investment would such traits persist. This constraint ensures that bioluminescence, though beautiful, would be highly functional. Alien light would be purposeful, shaped by ecological dynamics and evolutionary pressures, not merely aesthetic flourish.

Moreover, the spectral properties of this light would be shaped by the composition of the surrounding water and the biology of the emitter. On Earth, most deep-sea bioluminescence occurs in the blue-green part of the spectrum, as these wavelengths travel farthest in seawater. But on an alien world with different ocean chemistry—perhaps rich in ammonia or methane—the optimal wavelength might shift. This could lead to entirely different colour palettes: crimson pulses, violet flares, or infrared glows, invisible to the human eye. The ecology of such a world would be painted in hues we cannot currently imagine, shaped by alien constraints and innovations.

Understanding these possibilities has implications for astrobiology and exploration. If we are to one day detect life on distant worlds—especially in subsurface oceans—we must consider how it might signal its presence. Bioluminescence offers a tantalising, albeit indirect, possibility. Plumes venting from an icy crust might carry organic compounds or display fluctuating brightness. Future probes, whether orbiters, landers, or cryobots descending into subglacial seas, should be equipped not just to sample water and search for chemical signatures, but to observe darkness with sensitivity to unexpected light. In the vast night of alien oceans, a flicker could mean life.

Bioluminescent ecosystems also force us to confront deeper philosophical questions. If life can evolve in the absence of stars, generating its own light and constructing its own rhythms, then our anthropocentric view of life's conditions must be revised. Life does not wait for light—it creates it. It adapts not to comfort, but to opportunity. And in so doing, it reveals an essential truth: that existence is not dictated by environment alone, but by the capacity to transform environment into habitat.

In this way, the study of bioluminescence—on Earth and in theory—invites us to consider a more expansive view of the possible. It challenges us to abandon the comforting logic of sunlight and instead contemplate a universe of self-lit ecosystems, of creatures who pulse with internal energy, of worlds where night is not the absence of light but its source. It is a vision at once alien and familiar, strange yet grounded in evolutionary precedent.

Whether drifting through the cold void or locked beneath kilometres of ice, these potential ecosystems whisper a radical idea: that even in total darkness, evolution does not sleep. It adapts, innovates, and if conditions allow, it glows.

Gravitational Heating and Tidal Forces as Life Engines

When we think of the energy that powers life, our minds instinctively return to the Sun. Photosynthesis—arguably the most foundational biological process on Earth—anchors nearly all ecological systems in solar radiation. Yet beneath this planetary reliance on starlight lies a subtler truth: energy is everywhere, and in many forms. If life is a thermodynamic phenomenon that requires gradients—differences in temperature, chemistry, or pressure—then any mechanism that sustains such gradients over long timescales has the potential to drive biological activity. Among the most promising of these mechanisms are gravitational heating and tidal forces, subtle sculptors of planetary interiors that may be critical to the habitability of worlds shrouded in darkness.

The essence of tidal heating lies in gravitational flexing. When an orbiting body experiences changing gravitational forces—due to an elliptical orbit or the influence of neighbouring objects—its internal structure is periodically compressed and stretched. This deformation creates friction, and friction generates heat. Over time, this process can raise internal temperatures significantly, even in the absence of sunlight. If the body contains water, that heat may prevent it from freezing, enabling the persistence of subsurface oceans and hydrothermal activity. In certain configurations, tidal heating can become a dominant energy source—sufficient not only to keep oceans liquid, but to drive active geology, cryovolcanism, and potentially life-sustaining chemical cycles.

The most striking illustration of this phenomenon exists within our own solar system. Jupiter's moon Io is a volcanic powerhouse, the most geologically active body we know, driven entirely by tidal heating. As Io orbits Jupiter, its path is influenced by a gravitational tug-of-war involving its neighbours Europa and Ganymede. This constant gravitational interplay distorts Io's shape and generates immense internal friction. The resulting heat fuels hundreds of active volcanoes, vast lava lakes, and a continuously resurfacing crust. It is a violent example, but a telling one: starlight is not required for a planetary body to remain dynamically alive.

More germane to the search for life, however, are moons like Europa and Enceladus. Both are icy, both orbit gas giants, and both are believed to harbour subsurface oceans kept warm by tidal forces. On Europa, surface features such as ridges, chaos terrain, and linear fractures suggest a mobile ice shell, under which lies a global ocean tens of kilometres deep. The heat that maintains this ocean originates not from the Sun but from the rhythmic stretching induced by Jupiter's gravity. Similarly, Enceladus, though smaller than Europa, exhibits plumes of water vapour and organic compounds venting from fissures near its south pole. These geysers are thought to be the result of tidal stresses and subsurface pressure—an indication that liquid reservoirs, possibly connected to hydrothermal vents, exist beneath the moon's icy crust.

These examples reinforce a critical concept: the habitability of a world is not merely a matter of its position relative to a star. Rather, it depends on the internal energy budget of the planet or moon, and the ways in which that energy is distributed and maintained. Tidal heating allows for the decoupling of life from the so-called "habitable zone," expanding the field of potential biospheres beyond the traditional boundaries. In effect, it redefines the parameters of planetary habitability.

The mechanics of tidal heating depend on several factors: orbital eccentricity, the gravitational pull of nearby bodies, the elasticity of the planet's interior, and the presence of subsurface materials that can deform and generate friction. Worlds in highly elliptical orbits are particularly susceptible, as the varying gravitational forces across their orbit amplify the flexing process. Similarly, systems with resonant orbital relationships—where multiple moons orbit in integer ratios—can sustain long-term tidal interactions, preventing the orbit from circularising and the heating from dissipating.

This dynamic plays out dramatically in the Laplace resonance between Io, Europa, and Ganymede, where every orbit of Ganymede corresponds to two of Europa and four of Io. This synchronisation perpetuates orbital eccentricity, driving ongoing tidal friction. Were these moons isolated, their orbits would stabilise and their interiors cool. Instead, they remain energetically active—a state that may be mirrored in other planetary systems across the galaxy.

The implications for rogue planets and other non-traditional habitats are profound. Even a planet adrift in interstellar space could retain a source of tidal heating, provided it hosted a substantial moon. The gravitational interaction between the two bodies could generate enough internal heat to support liquid water, hydrothermal activity, and potentially biology beneath an icy crust. Moreover, such interactions could persist for billions of years, offering stability over timescales conducive to evolutionary development.

Beyond moons and rogue planets, tidal heating may also operate within multi-planet systems around low-mass stars. M-dwarf stars, though faint, often possess tightly packed planetary systems. In such configurations, gravitational resonances between planets could generate internal heating on a planetary scale. These systems are especially intriguing because their stars are long-lived and abundant, making them prime targets for exoplanet studies. While many of these planets orbit close to their host stars, potentially exposing them to harmful radiation, internal heating mechanisms could support subsurface environments shielded by thick crusts or atmospheres.

Gravitational heating is also significant in the context of exoplanetary atmospheric dynamics. On planets with internal oceans or magma layers, the heat generated from tidal forces can influence surface geology and atmospheric composition. Volcanic outgassing, for instance, could replenish a planet's atmosphere with greenhouse gases, contributing to climate stability. On Earth, volcanism and tectonic activity are crucial components of the carbon cycle—a key regulator of temperature over geological timescales. In similar fashion, tidal heating on exoplanets might support feedback systems that extend the lifespan of habitable conditions, even under weak stellar illumination.

From a biosignature detection perspective, gravitational heating introduces complexity and opportunity. Traditional biosignatures, such as oxygen or methane, are often tied to surface processes and photosynthesis. However, on tidally heated worlds, the indicators of life may be more subtle—plumes of water vapour, transient atmospheric components, or anomalous heat emissions. Understanding the thermal and chemical fingerprints of these processes will be essential as we develop more sensitive instruments and refine our models of habitability.

There are, of course, limitations. Tidal heating is not inexhaustible. Over time, orbital interactions can stabilise, and the heat can wane. A body without replenishment—either through continued gravitational resonance or radioactive decay—will eventually cool. Moreover, excessive tidal heating can be destructive, as seen with Io, where extreme volcanism may preclude the stability needed for life. The balance must be just right: enough energy to sustain liquid water and geological activity, but not so much as to render the environment hostile or volatile.

Nevertheless, gravitational heating and tidal forces represent one of the most promising avenues for expanding the scope of planetary habitability. They highlight the fact that life need not be anchored to a sunbeam. It can thrive in the deep, warm recesses of ice-covered oceans, in the volcanic pulses of subterranean worlds, and in the slow, rhythmic breathing of planets caught in gravitational resonance. They reveal that energy is not a privilege of proximity, but a function of interaction—that in a universe bound by gravity, warmth may arise in the most unexpected places.

As we map the contours of this new scientific frontier, we are also reshaping our expectations. The quiet interiors of moons and planets—once thought to be cold and dead—may in fact be alive with motion, chemistry, and the subtle potential for life. In their heating, they echo a deeper truth: that the cosmos, vast and indifferent though it may seem, offers not a single mould for life, but a diversity of conditions through which life might emerge, adapt, and endure. It is up to us now, with sharpened instruments and broadened imaginations, to follow that heat wherever it leads.

Psychological Implications of Lifeless Starscapes

Among the most haunting concepts in modern astronomy is that of the rogue planet—a world untethered, drifting alone through the void, orphaned from any stellar warmth or gravitational home. These dark wanderers evoke a peculiar blend of fascination and disquiet, challenging not only our scientific understanding of planetary formation, but also stirring deep emotional and existential responses. If the cosmos reflects our inner world as much as it reveals an external reality, then rogue planets—lifeless starscapes in the truest sense—hold up a profound psychological mirror. Their silence is not merely physical. It resonates within us, conjuring themes of exile, abandonment, and the delicate tension between solitude and connection.

The sheer image of a rogue planet is enough to unsettle even the most seasoned thinker. Picture it: a vast, frozen sphere adrift in blackness, absent of day or night, its sky forever devoid of sunrise, its orbit not a path but a departure. There are no seasons to mark the passage of time, no shadows to give structure, no gravitational centre to lend stability. It is the ultimate in cosmic solitude. And yet, we now believe these worlds may be common—perhaps as numerous as the stars themselves. That realisation, while scientifically exhilarating, is also psychologically disconcerting. For if the universe abounds with isolated planets, it may also suggest that isolation is not an aberration but a feature of existence.

Human consciousness, shaped as it is by sociality and context, instinctively recoils from such loneliness. We are, by nature and necessity, relational beings—bound by ties to others, to places, and to narratives. From early childhood, our psychological health is contingent on connection: to parents, to community, to meaning. The concept of an untethered world therefore unsettles us at a symbolic level. It challenges our assumption that life, like planets, must revolve around something—that identity and purpose are necessarily grounded in proximity to warmth, whether emotional or stellar.

This sense of dislocation is not merely poetic. It has concrete psychological analogues. Clinical psychologists often describe existential isolation as one of the four fundamental concerns of human experience, alongside death, freedom, and meaninglessness. Rogue planets offer a literal embodiment of this isolation—worlds adrift in a contextless void. In psychoanalytic terms, they might be seen as the unconscious projection of disintegrated selves—those fragments of identity that have lost their centre, their anchoring points in the cosmos of the psyche. They are the dissociated memories, the unmet needs, the exiled parts of our inner worlds, floating unseen in the background of conscious life.

Moreover, the metaphor of the rogue planet resonates with cultural and literary archetypes. In myth and literature, exile is a recurring theme—the castaway, the wanderer, the outcast. These figures often reflect a psychological state of detachment or alienation. Odysseus, the Wandering Jew, Frankenstein's creature, even the modern astronaut—each embodies a condition of liminality, existing on the margins of community and cosmos. Rogue planets are the astronomical counterparts of these myths. They are not merely astronomical phenomena, but narrative entities—symbols of otherness and the fear of not belonging.

The psychological weight of this symbolism becomes even more apparent when placed in the context of the Fermi Paradox. If intelligent life is common, why have we found no evidence of it?

Some theorists propose that advanced civilisations may destroy themselves before achieving interstellar communication. Others suggest that life is rare, or that the distances are too great. But another, more introspective possibility is that life does exist, and that it too experiences the silence of cosmic isolation. The rogue planet, in this context, becomes a model for not just physical, but communicative loneliness. What if we are not alone, but unable to reach? What if all life, like rogue worlds, is locked in solitary drift?

This perspective invites a confrontation with the limits of human agency and the narratives we construct to give coherence to our existence. The Copernican revolution dethroned Earth from the centre of the universe. The discovery of exoplanets dethroned our solar system as unique. The realisation that some planets exist without stars—without warmth, without orbit, without illumination—may similarly destabilise our existential assumptions. It suggests that coherence is not guaranteed, that anchoring is not universal, and that perhaps not all worlds—and not all lives—have a centre.

Yet, in the face of this unsettling possibility, a counter-theme emerges: the resilience of life and meaning even in the absence of familiar frameworks. If rogue planets can, through internal heat and chemical processes, sustain subsurface oceans—and potentially life—then so too can isolated individuals find inner resources for survival, creativity, and connection. The rogue planet becomes not just a symbol of exile, but of autonomy and interiority. It embodies the idea that warmth can come from within, that illumination need not be external, and that life, even if hidden, can persist.

This reframing has psychological relevance. In therapeutic contexts, individuals often grapple with feelings of disconnection—estrangement from family, loss of community, spiritual disorientation. The metaphor of the rogue planet, bleak as it may seem, also holds a quiet hope: that isolation is not the end of the story, but a phase, a state from which new forms of integration may emerge. The darkness need not mean death; it may be gestational. The absence of orbit does not imply purposelessness—it may signal a transition to a new kind of self-defined path.

The emotional valence of rogue planets is further heightened by the aesthetics of space itself. The vastness of interstellar voids, the silence of lightless regions, the slow procession of time in gravitationally detached settings—all provoke a unique psychological response. Astronauts and cosmonauts, despite orbiting well within our star's reach, report profound experiences of awe, insignificance, and sometimes ontological panic. This "overview effect" reveals that even proximity to Earth cannot fully shield the psyche from the weight of cosmic scale. How much more potent, then, would the psychological impact be on a sentient being native to or stranded on a rogue world?

Speculative fiction has begun to explore such questions, often portraying rogue planets as settings for dystopia, introspection, or transcendence. Whether as metaphors for ecological collapse, societal alienation, or metaphysical solitude, these lifeless starscapes allow storytellers to examine what it means to live without light. They also challenge the boundaries of empathy. If we can imagine life evolving—or continuing—on a world without a sun, can we also extend compassion to the parts of ourselves and others that live in darkness, outside the orbit of normative experience?

Rogue planets, then, are not only testaments to astronomical curiosity. They are psychological artefacts—reminders of our fascination with solitude, our fear of abandonment, and our longing for reconnection. They represent both exile and autonomy, desolation and durability. In their quiet drift, they echo the human condition in its most reflective state: searching not for light, but for meaning in its absence.

Ultimately, the contemplation of lifeless starscapes does not have to end in despair. It can open a space for new metaphors, new models of selfhood, and new forms of solidarity. Just as rogue planets may carry the seeds of life within their buried oceans, so too can moments of psychological darkness hold potential for regeneration. In this way, astronomy becomes not just a study of what is out there, but a deeper exploration of what is within. The stars may guide us, but it is often in their absence that we come to understand what we truly seek.

Chapter 4
Biosignatures and the Problem of False Positives

To search for life beyond Earth is to pursue the most tantalising of scientific goals—one that dances at the intersection of astronomy, chemistry, biology, and philosophy. It is a pursuit both thrilling and sobering, for it requires not only the courage to look outward but the discipline to interpret uncertainty with restraint. At its core lies the quest to identify biosignatures: measurable characteristics in the atmosphere or on the surface of an exoplanet that could plausibly indicate the presence of life. Yet the more sophisticated our tools become, the clearer it is that the cosmos is a master of ambiguity. Many of the signs we long for—oxygen, methane, unusual chemical imbalances—can also be generated by lifeless processes. The risk is not merely in failing to detect life, but in prematurely declaring its presence. The history of biosignature science is, in many ways, a history of false positives and hopeful misinterpretations.

This chapter delves into that critical tension—the precarious space between detection and declaration, between a tantalising signal and a definitive claim. It begins with a foundational question: what truly counts as evidence for life? At first glance, the answer seems straightforward. On Earth, life modifies its environment in detectable ways. Photosynthesis releases oxygen, methanogens produce methane, and biological activity often maintains atmospheric compositions far from chemical equilibrium. Therefore, if we were to find a planet with high levels of oxygen and methane coexisting, we might reasonably infer a biological source. But nature, complex and creatively indifferent, does not cater to our inferences. Abiotic processes—such as photodissociation, volcanic outgassing, and surface-atmosphere interactions—can produce similar signatures. The line between biology and chemistry is porous, and without careful context, we risk mistaking the latter for the former.

Nowhere has this ambiguity been more evident than in the case of Mars. For decades, methane detections on the Red Planet have sparked intense debate. Methane is a fleeting molecule; it breaks down relatively quickly under solar radiation, which means any detected quantity must have been recently produced. Its presence, therefore, is intriguing. On Earth, methane is largely biogenic. But on Mars, the sources are unclear. Could it be geological—emerging from serpentinisation reactions within the crust? Could it be the product of ancient organics released by seasonal heating? Or could it be the residue of microbial life beneath the surface, as some have hoped? The answer remains elusive, not least because different missions have reported conflicting results. The European Space Agency's Trace Gas Orbiter found little to no methane, even as the Curiosity rover registered transient plumes. It is a perfect illustration of how data alone do not yield conclusions—interpretation is shaped by assumptions, models, and sometimes, unconscious desire.

Exoplanets present an even more formidable challenge. The detection of oxygen in an alien atmosphere would undoubtedly make headlines, yet we must ask: does oxygen always mean life? In Earth's history, oxygen was not a dominant atmospheric component until around 2.4 billion years ago, following the Great Oxygenation Event driven by cyanobacteria. Before that, life thrived in anaerobic conditions. Moreover, oxygen can be produced abiotically through the photodissociation of water vapour under high ultraviolet flux, especially around M-dwarf stars. Similarly, methane can result from volcanic activity or cometary delivery. Even more complex molecules—such as chloromethane or ethylene oxide—can have non-biological origins under certain conditions. Thus, the interpretation of biosignatures is never a matter of detecting a molecule in isolation, but of situating that molecule within a planetary, stellar, and temporal context.

Adding to this complexity is the role of instrumental noise and cognitive bias. Every detection involves a chain of inference—an instrument records a signal, which is translated into a spectrum, which is modelled against templates, which are interpreted by scientists. At each stage, error can creep in. Telescopes have limited resolution; stellar activity can mimic or mask spectral features;

atmospheric models are simplifications of reality. Moreover, scientists, like all humans, bring expectations to their data. Pattern recognition, while a powerful cognitive tool, can lead to overfitting—finding significance where none exists. The desire to detect life is profound and sincere, but it can also distort. In this context, rigorous scepticism is not cynicism but a safeguard.

Even more provocatively, some thinkers have entertained the idea that alien life—if it exists—might actively evade detection. The possibility that life could mimic abiotic processes or exist in forms so unfamiliar that we overlook them raises unsettling questions. Could life conceal itself within chemical noise? Could its biosignatures resemble geochemical processes so closely that we misclassify them? These are not idle fantasies. On Earth, countless organisms camouflage themselves from predators and prey alike; mimicry is a well-established evolutionary strategy. If life exists elsewhere and has undergone natural selection, it is not inconceivable that it might have developed analogous behaviours. Alternatively, it may not be hiding at all, but rather it may be so fundamentally different from life as we know it—perhaps based on silicon, ammonia, or entirely novel biochemistries—that our instruments, and more importantly our minds, fail to register it.

All of this points to the need for a more robust, multidisciplinary approach to biosignature detection—one that goes beyond individual molecules and considers broader system-level patterns. One promising avenue is the application of systems theory, which focuses on the interactions and feedbacks among components of a system. A planet with life may exhibit dynamic regulation—such as the Gaia hypothesis suggests—maintaining homeostasis through biological-environmental coupling. Instead of single-gas detection, we could look for complex atmospheric disequilibria that require continual energy input to maintain. Other proposals include using entropy-based metrics to assess whether a planetary system exhibits structures or flows indicative of active information processing or thermodynamic optimisation—signs that may suggest the presence of a metabolic network or even rudimentary intelligence.

Furthermore, developments in artificial intelligence and machine learning could play a crucial role in expanding our detection capabilities. Pattern recognition algorithms, trained not just on Earth-based templates but on a wide array of simulated environments, may be able to identify subtle or unexpected signatures that human analysts might miss. However, this approach too must be tempered with caution, as AI systems are only as good as the data and assumptions on which they are trained. Teaching a machine to "imagine the unknown" remains an unsolved challenge.

The epistemological humility required in this field is significant. The temptation to make definitive claims is strong, especially in a media landscape hungry for discovery. But science advances not by proclamations but by the gradual accumulation of evidence, sceptical review, and the careful refinement of theory. The detection of life will not come from a single molecule or even a single planet. It will emerge from the convergence of multiple lines of evidence, from anomalies that persist across contexts, from patterns that defy abiotic explanation, and from signals that remain coherent under scrutiny.

In this chapter, we will dissect the current state of biosignature research: the biomarkers under investigation, the historical case studies that illustrate the perils of premature interpretation, the instrumental limitations that constrain our confidence, and the philosophical frameworks that might offer a more nuanced path forward. We will explore the idea that life, if it is to be recognised beyond Earth, must be approached not only with tools of science but with the patience of philosophy, the imagination of art, and the rigour of sceptical reasoning.

For in the end, the problem of false positives is not just about technical errors or ambiguous molecules. It is about the human longing to see ourselves reflected in the cosmos—to find, in the chemical haze of a distant world, the echo of a living breath. And while that longing is noble, it must be matched by the discipline to ask: are we truly seeing life, or merely projecting it? Only by wrestling honestly with that question can we prepare ourselves for the day—if it comes—when the signal is clear, the context sound, and the evidence undeniable. Until then, the search must continue—not as a rush to affirmation, but as a careful cultivation of knowledge, wonder, and humility.

What Counts as Evidence for Life?

The search for extraterrestrial life has never been a straightforward scientific pursuit. It is a philosophical endeavour veiled in empirical rigour, a venture propelled as much by human curiosity as by chemical analysis. At its heart lies a deceptively complex question: what counts as evidence for life? It is a query that demands not only an inventory of molecules and measurements but a re-examination of our assumptions—both biological and epistemological. As our tools grow more sensitive and our catalogue of exoplanets swells, the need for precision, caution, and conceptual clarity becomes paramount. To avoid mistaking wishful thinking for discovery, we must critically assess the biomarkers we pursue and understand why many of them are not, in themselves, conclusive.

In the field of astrobiology, a biosignature is broadly defined as any substance—or pattern of substances—that provides scientific evidence of past or present life. These signatures may be gases in a planet's atmosphere, isotopic ratios in surface materials, or complex organic molecules in the dust of comets. But the most discussed—and most scrutinised—are atmospheric biomarkers: gases that life tends to produce and that persist at detectable levels due to biological replenishment. On Earth, the leading candidates include oxygen, ozone, methane, nitrous oxide, and combinations of gases that together suggest a state of chemical disequilibrium.

Oxygen, for example, is often cited as the quintessential biosignature. Its abundance in Earth's atmosphere is maintained by photosynthetic organisms that split water molecules and release oxygen as a byproduct. Without constant biological input, atmospheric oxygen would quickly react with surface minerals and disappear. From this, we infer that a high concentration of oxygen on an exoplanet might signal biological activity. And yet, there are ways for oxygen to accumulate abiotically. Photodissociation of water vapour under intense ultraviolet light can release oxygen atoms, which then recombine into O_2 molecules. This process is especially likely around M-dwarf stars, which emit strong UV radiation and often have planets close enough to be tidally locked—conditions conducive to both water loss and oxygen build-up. In such scenarios, a seemingly life-indicative signal may in fact be the product of photochemistry alone.

Methane presents a similar conundrum. On Earth, most atmospheric methane arises from biological sources—rice paddies, wetlands, and the digestive tracts of ruminants. However, methane can also be generated geologically through serpentinisation, a reaction between water and olivine-rich rocks in the mantle. This process produces hydrogen, which can then reduce carbon dioxide to methane. The detection of methane on Mars, therefore, has triggered both excitement and scepticism. Observations by the Curiosity rover have shown seasonal variations in methane concentrations, but without an accompanying rise in ethane or propane—gases typically associated with biotic methane—the case remains unresolved. The ambiguity of the signal has only been compounded by conflicting data from orbiters, some of which have failed to detect the gas altogether. Methane, it seems, is a prime example of how a molecule can be deeply evocative without being conclusive.

This raises an important methodological challenge: biosignatures are not binary indicators. They are probabilistic signals, whose interpretation depends on context. A gas that suggests life in one environment might be irrelevant or misleading in another. To assess a biosignature's validity, we must consider planetary temperature, pressure, stellar radiation, chemical cycles, and geological activity. Only when these variables are accounted for can we estimate whether a given molecule is likely to be biological in origin.

The complexity increases further when we look at combinations of gases. One proposed strategy is to identify atmospheres in disequilibrium—conditions that, if left to natural chemistry, would not persist. For example, the coexistence of oxygen and methane in significant quantities is considered unlikely in a purely abiotic system, since the two gases readily react with one another. Their simultaneous presence implies continuous replenishment by incompatible sources, which may suggest metabolic activity. Yet even this approach is not foolproof. Certain abiotic processes, such as volcanic eruptions coupled with photochemistry, could theoretically maintain a state of temporary disequilibrium. Moreover, our models are Earth-centric. What appears unstable from our perspective might be sustainable under unfamiliar thermodynamic or kinetic constraints.

The problem is not limited to atmospheric chemistry. Isotopic ratios—often touted as signs of life—can also yield false positives. Biological processes preferentially incorporate lighter isotopes, such as carbon-12 over carbon-13. This "fractionation" leads to isotope ratios that deviate from those expected in abiotic systems. The famous case of the Martian meteorite ALH84001, discovered in Antarctica, brought this issue into sharp focus. In 1996, researchers announced the presence of carbonate globules and potential microbial fossils, along with isotopic anomalies suggestive of biological activity. The findings made international headlines. But subsequent analyses revealed that the structures could be explained by inorganic processes, and the isotope ratios, while unusual, were not definitive. The enthusiasm outpaced the evidence, and the episode now serves as a cautionary tale in scientific interpretation.

Beyond the technical and chemical dimensions, there is also the psychological component to consider. Scientists, like all humans, are prone to pattern recognition and confirmation bias. In a field where data are sparse and stakes are high, the temptation to see life where there is none is powerful. Spectral features can be noisy, ambiguous, or artefactual. Instruments have limits, and interpretation is always nested within theoretical frameworks. This makes rigorous scepticism essential. To claim a biosignature is to imply not only the detection of a molecule, but the exclusion of all plausible abiotic explanations. It requires a high standard of evidence—one that few signals have yet reached.

This does not mean that the search is futile, but that our criteria must evolve. Increasingly, astrobiologists are moving toward a systems-level understanding of biosignatures. Rather than hunting for individual molecules, researchers are examining broader planetary processes: redox balances, surface-atmosphere interactions, and global energy flows. The goal is to discern patterns of behaviour—feedback loops, persistent disequilibria, thermal anomalies—that cannot be easily explained by abiotic models. Such an approach, while more complex, offers a richer and potentially more reliable framework for life detection.

Equally important is the need to diversify our assumptions. Life as we know it is based on carbon, water, and moderate temperatures. But this may be a parochial view. Silicon-based biochemistries, ammonia solvents, or entirely novel forms of metabolism may exist elsewhere. These life forms may not emit oxygen or methane; they may not even modify their environments in detectable ways. To find them, we must be willing to search not just for Earth 2.0, but for life 2.0—a reimagined, open-ended concept that respects the creative possibilities of evolution under alien conditions.

In practice, this will require both theoretical innovation and technological advancement. Future missions must be equipped not only with spectrometers and chromatographs, but with algorithms capable of recognising the unexpected. We must integrate insights from biology, chemistry, planetary science, and information theory to construct a robust framework for biosignature interpretation. And we must cultivate a culture of humility—a recognition that the cosmos may not present its answers in forms we are prepared to receive.

What counts as evidence for life, then, is not a static list of biomarkers. It is a dynamic, context-dependent evaluation—a synthesis of signal, background, environment, and inference. It is a question not only of molecules, but of models. Not only of detection, but of interpretation. And not only of science, but of the values and expectations that shape our encounter with the unknown. In facing this question honestly, we do not diminish the excitement of discovery; we enhance its credibility. For when the signal finally comes—clear, repeated, and irreducibly biological—it will stand not as the triumph of a single molecule, but as the outcome of careful thought, disciplined scepticism, and an enduring respect for complexity. Only then will we know not just what we've seen, but what it means.

Methane on Mars, Oxygen on Exoplanets: A Cosmic Rorschach Test

In the delicate and often speculative art of astrobiology, few substances have generated more fervour and debate than methane and oxygen. These gases, familiar from Earth's biosphere, carry the weight of implication: they hint at metabolism, at photosynthesis, at processes we instinctively associate with life. Yet their presence elsewhere—on barren Martian plains or in the alien skies of exoplanets—has proven to be as much a source of ambiguity as excitement. In the search for extraterrestrial life, both methane and oxygen have become Rorschach tests for the scientific imagination. Their detection prompts not only technical analysis but deeper psychological reflection on our interpretive frameworks, our hopes, and our blind spots.

Nowhere is this interpretive tension more evident than in the ongoing saga of Martian methane. The Red Planet, long a focal point of life-related speculation, continues to offer just enough intrigue to resist dismissal, yet never enough certainty to confirm biological activity. Methane has been observed on Mars in transient, fluctuating quantities by multiple instruments over the past two decades. The most prominent of these observations came from NASA's Curiosity rover, which reported seasonal variations in methane concentration at its landing site in Gale Crater. Levels peaked during the Martian summer and diminished during winter—a pattern suggestive of dynamic, possibly cyclical processes.

The implications were profound. On Earth, over 90% of atmospheric methane is biological in origin, produced by microbial life, ruminant digestion, and decaying organic material. If Mars exhibits a similar signature, it could indicate ongoing subsurface biology. However, the chemical ambiguity of methane undermines this hypothesis. Serpentinisation—a geochemical process in which water reacts with certain iron-rich rocks—can also generate methane without life. The Martian crust, with its known olivine deposits, provides a plausible context for such reactions, particularly if residual heat or aqueous environments persist below the surface.

Further complicating the picture is the inconsistency of detection. The European Space Agency's Trace Gas Orbiter (TGO), a mission specifically designed to analyse atmospheric gases, reported an absence of methane at the levels observed by Curiosity. Its sensors, operating from orbit, should have detected the plumes with ease—yet they did not. This discrepancy has sparked considerable debate. Are the methane emissions hyper-localised and short-lived, escaping detection from above? Is there a yet-undiscovered sink that rapidly scrubs methane from the atmosphere? Or are the instruments, calibrated differently and operating in separate environments, producing irreconcilable data sets?

These unresolved questions illustrate the core challenge of interpreting biosignatures: even well-studied molecules can mislead when isolated from context. Methane on Mars remains enigmatic not because we lack data, but because we lack a unifying theory to make sense of the data. Each new detection is a Rorschach blot—inviting different interpretations depending on one's scientific perspective, disciplinary background, or cognitive bias. The gas itself is silent; it is our frameworks that give it voice.

A similar interpretive dance plays out in the realm of exoplanetary science, where the detection of oxygen in distant atmospheres carries heavy symbolic weight. On Earth, oxygen is both a product and a prerequisite of complex life. Its abundance is a direct consequence of photosynthetic activity by cyanobacteria and, later, plants. Detecting oxygen elsewhere, especially in combination with reducing gases like methane, would be a major milestone in the search for extraterrestrial biology. It would suggest not just habitability, but habitation.

And yet, as with Martian methane, the story is more complicated. Oxygen can accumulate abiotically under certain conditions. In particular, planets orbiting M-dwarf stars—small, cool stars that make up the bulk of the galaxy's stellar population—are subject to intense ultraviolet radiation. This radiation can split water vapour molecules in the upper atmosphere, allowing hydrogen to escape into space and leaving oxygen behind. Over time, this process can produce an oxygen-rich atmosphere in the absence of life. Such "false positive" scenarios are now a central concern for exoplanet researchers, prompting more nuanced interpretations of spectral data.

In one striking example, the exoplanet candidate LHS 1140 b—a super-Earth in the habitable zone of a nearby M-dwarf—sparked interest for its potential atmospheric composition. Some models suggested that the planet could possess an oxygen-dominated atmosphere as a result of runaway water loss. Yet such a biosignature, while chemically suggestive, would not necessarily reflect biology. Instead, it might represent planetary desiccation—a process that leaves oxygen as a kind of geochemical relic, not a metabolic byproduct. Here again, context is everything: the same molecule that signifies life on Earth might signify its absence elsewhere.

The metaphor of the Rorschach test is especially apt because it highlights the role of interpretation in scientific inquiry. Biosignatures do not speak for themselves. They are filtered through theoretical models, instrumental limitations, and the psychology of discovery. Scientists, like all humans, are prone to confirmation bias—the tendency to interpret ambiguous data in ways that affirm prior beliefs. The desire to find life is powerful and deeply embedded in our cultural narrative. It influences the questions we ask, the frameworks we build, and the conclusions we draw. This is not a flaw, but a feature of human cognition. The challenge is to make it visible and accountable within the scientific process.

Some researchers have responded to these complexities by advocating for multi-dimensional detection frameworks. Rather than relying on single-molecule biomarkers, these approaches seek patterns of co-occurring signals that defy abiotic explanation. For instance, the simultaneous presence of methane, oxygen, water vapour, and certain organic compounds might collectively suggest biological processes, especially if the atmospheric composition remains out of equilibrium. But even this strategy is probabilistic, not deterministic. It strengthens the inference of life, but it does not confirm it.

Others have turned to systems theory, entropy analysis, and the search for complex feedback loops that may indicate life's influence on a planetary environment. These approaches aim to shift the emphasis from specific chemicals to broader patterns of interaction and regulation. They are promising but also speculative, requiring new models, new tools, and perhaps new ways of thinking about what life is and how it manifests.

Ultimately, the ambiguity surrounding methane on Mars and oxygen on exoplanets serves as a humbling reminder of the epistemic limits of science at the frontier. Our instruments grow more powerful, our data sets more detailed, and our models more sophisticated. Yet the core question—are we alone?—remains unanswered not for lack of effort, but because nature is complex, subtle, and often indifferent to our categories. It is not enough to detect a molecule. We must also understand the context in which it exists, the processes that might produce it, and the theoretical scaffolding through which we interpret it.

To navigate this uncertainty, we must cultivate a blend of scientific rigour and philosophical openness. We must be willing to hold multiple hypotheses in tension, to revise our assumptions in light of new evidence, and to resist the allure of premature conclusions. For while the search for life beyond Earth is a scientific enterprise, it is also a human one—driven by curiosity, shaped by imagination, and sustained by the hope that, somewhere in the cosmos, we are not alone.

In that hope, methane and oxygen are not just molecules. They are symbols—powerful, ambiguous, and endlessly interpreted. Like inkblots on the canvas of the universe, they reveal not only what might be out there, but what we are prepared to see.

Detecting the Undetectable: Noise, Bias, and the Illusion of Certainty

In the great enterprise of searching for life beyond Earth, perhaps the most dangerous illusion is that of clarity. A sharply defined spectral line, a bump in a light curve, a spike in chemical concentration—these data points can seduce us into believing we have glimpsed something definitive. Yet behind every apparent signal lies a layered architecture of assumptions, calibrations, statistical models, and human interpretation. As we refine our instruments to search the skies with ever-increasing sensitivity, we also enter a realm where the margin between signal and noise narrows dramatically. At this frontier, the greatest challenge is not always detecting new phenomena—it is knowing when we have actually detected them at all.

The limits of our instrumentation are, by their nature, both technical and conceptual. No measurement device is perfect; each has thresholds of detection, sources of error, and a relationship to the physical world mediated by calibration and theory. Consider the example of transit spectroscopy, one of the primary methods for assessing exoplanet atmospheres. When a planet crosses in front of its host star, some of the starlight passes through the planet's atmosphere. Molecules in that atmosphere absorb light at specific wavelengths, producing tiny dips in the spectrum we observe. These dips are often no greater than a few dozen parts per million—a degree of subtlety that demands not only exquisite instrumentation but equally rigorous interpretation.

Yet these same instruments must contend with a host of complications. Stellar activity—flares, spots, and pulsations—can mimic or obscure the signals we hope to extract. Instrumental drift, cosmic ray hits, and thermal fluctuations can introduce artefacts. Even the process of removing these known distortions is itself susceptible to bias. Choices made during data processing—how to detrend a light curve, which noise model to apply, how to fit a baseline—can dramatically affect the results. A well-known cautionary tale is the debated detection of phosphine in the atmosphere of Venus in 2020. Initial reports suggested a signature consistent with this rare molecule, which on Earth is almost exclusively associated with anaerobic life. The findings ignited global excitement. But subsequent re-analyses cast doubt on the detection, attributing the signal to background calibration errors, spectral confusion with sulphur dioxide, or assumptions embedded in the data reduction pipeline. The episode underscored how fragile conclusions can be when drawn from signals so close to the noise floor.

The Venus phosphine controversy is instructive not only for its technical missteps but for the cognitive environment in which it unfolded. Scientists are not immune to bias, particularly when interpreting ambiguous data in a field charged with existential implications. Confirmation bias—the tendency to favour information that supports our pre-existing beliefs—can skew interpretation even in the most well-controlled environments. This is not a character flaw; it is a feature of human cognition. In an arena like astrobiology, where the data are sparse and the stakes high, this cognitive tendency becomes especially pronounced.

Moreover, statistical methods themselves can become complicit in this illusion of certainty. Bayesian inference, machine learning, and other tools of modern data science are powerful, but they are not oracles. They rely on prior assumptions and model structures that can subtly influence outcomes. For example, if one assumes that life is common, one might be more inclined to interpret a marginal biosignature as meaningful. If, conversely, one assumes life is vanishingly rare, one may dismiss the same signal as noise. The data do not exist in a vacuum; they are processed through probabilistic lenses that often encode philosophical positions as much as empirical ones.

Even in seemingly straightforward measurements, the line between detection and artefact can be blurry. In the case of methane on Mars, varying readings from different missions—including Curiosity, the Mars Express orbiter, and the Trace Gas Orbiter—have offered conflicting results. Some recorded transient methane plumes; others saw nothing. This inconsistency is not necessarily evidence of error, but it reveals how localised, episodic, and complex the phenomenon might be. It also demonstrates the perils of overinterpreting any single dataset in isolation. Each instrument has its own design, altitude, sensitivity, and analytical pipeline. Cross-validation between instruments is critical, yet difficult, particularly when the data themselves are at the edge of detectability.

Beyond the technical and statistical dimensions lies an even subtler challenge: our metaphors and mental models. We speak of "signatures," "biosignals," "detected molecules," as though these were direct observations. But what we are often interpreting are differences in pixel intensities, shifts in spectral baselines, or deviations from expected curves. These are proxies—translated through layers of modelling and assumption. The risk is that we reify abstractions into objects, treating the outcome of a noisy statistical inference as if it were an unambiguous discovery. This is not to disparage the science; it is to advocate for a greater epistemic humility.

One practical response to these challenges has been the increasing use of simulation in support of observational astronomy. Teams now generate synthetic spectra of exoplanet atmospheres under a wide range of physical conditions to test whether certain biosignatures could plausibly arise without life. These simulations are not just diagnostic tools—they are philosophical instruments, forcing us to confront the conditional nature of our conclusions. Could an oxygen-rich atmosphere arise from photodissociation? Could methane persist without biology under certain geochemical constraints? Could a combination of both be sustained in equilibrium only by exotic abiotic mechanisms? These are not rhetorical questions—they are testable hypotheses, provided our models are transparent and our assumptions scrutinised.

Yet even the most robust models must ultimately contend with the reality of imperfect knowledge. The further we push the boundary of detectability, the more we must live with ambiguity. A spike in methane here, a spectral feature there—none of these alone constitute proof. The illusion of certainty is dangerous not because it leads us astray, but because it closes off the space for uncertainty, which is where genuine discovery occurs. To mistake a tentative signal for a definitive answer is to risk eroding the credibility of the very science we seek to advance.

This is especially important when communicating findings to the public. The media, driven by attention and simplification, often translates nuanced scientific claims into declarative headlines. A molecule is detected—it must be life. A planet lies in the habitable zone—it must be habitable. These narratives, while understandable, flatten the epistemic gradient that underpins scientific inquiry. They turn probabilities into certainties, and signals into stories. Scientists, therefore, bear a responsibility not only to detect and interpret, but to communicate with precision and restraint.

Ultimately, the challenge of detecting the undetectable is not a technological one alone. It is a philosophical, psychological, and linguistic challenge. It demands instruments that are not only finely tuned, but honestly calibrated against the limits of inference. It demands scientists who are not only experts in their fields, but aware of their cognitive biases and methodological assumptions. And it demands a community—scientific and public alike—that values not just discovery, but the rigour, humility, and openness that make discovery meaningful.

The cosmos is under no obligation to be clear. Its signals are faint, its processes complex, and its mysteries stubbornly layered. But in wrestling with these ambiguities, we refine not just our instruments, but our understanding of what it means to know. We learn to see not just what we expect, but what is there—however faintly, however fleetingly. And in that effort, we come a little closer to the truth—not by claiming certainty, but by learning to live, and think, within the necessary confines of doubt.

Life That Lies: Mimicry, Camouflage, and Alien Deception

In the search for extraterrestrial life, one of the most enduring assumptions has been that if alien organisms exist, we will eventually detect them—that life, once present and sufficiently active, will reveal itself through chemical footprints, thermal patterns, or electromagnetic signals. Yet this assumption may itself be a profound limitation. What if life is not merely elusive but actively disguised—intentionally or unintentionally concealed through biological mimicry, environmental camouflage, or technological deception? What if alien life doesn't just avoid detection but resists it, either as a byproduct of its evolutionary history or as a strategic choice? In contemplating this possibility, we are forced to confront not only the outer edge of scientific speculation but also deeper philosophical and ethical questions about recognition, misunderstanding, and the boundaries of what we are willing—or able—to see.

The concept of camouflage and mimicry is deeply familiar from terrestrial biology. On Earth, organisms routinely evolve to hide in plain sight. The peppered moth, whose colouration adapted to industrial soot, the leaf-tailed gecko that vanishes against bark, and the octopus that shifts shape and texture to blend with coral reefs—all represent sophisticated evolutionary strategies for avoiding detection by predators or prey. Mimicry extends further: the harmless king snake mimics the venomous coral snake's colouration, while orchids imitate female insects to lure pollinators. These are not marginal anomalies but mainstream strategies in nature's evolutionary arsenal. The capacity to deceive—to mislead other organisms—is a marker not of malicious intent, but of adaptive intelligence.

It would be parochial to assume such strategies are confined to Earth. If life exists elsewhere, it may be subject to similar selective pressures. Alien ecosystems could favour concealment as much as display. In such scenarios, the absence of expected biosignatures might not indicate the absence of life, but its success in hiding. We are, after all, only searching for life as we expect it to be—active, noisy, chemically expressive. What if alien life is metabolically dormant, chemically minimal, or cloaked within abiotic shells? What if it has learned, through its own evolutionary trials, that visibility invites danger?

This possibility takes on added significance when we consider the asymmetry of our detection technologies. We scan the cosmos for oxygen, methane, chlorophyll-like pigments, or artificial electromagnetic emissions. But each of these is predicated on assumptions about life's chemistry and communication methods—assumptions drawn entirely from a single biosphere. Should an alien species possess radically different biochemistry—silicon-based, ammonia-solvent, plasma-structured—it may generate no biosignatures that overlap with our detection protocols. Its chemical presence might be indistinguishable from geological background noise. Or, more intriguingly, it might mimic geophysical signals so effectively that we misclassify its biosphere as barren rock or atmospheric haze.

Some thinkers have extended this hypothesis into even more provocative territory: what if some alien life deliberately avoids detection—not merely as an evolutionary byproduct, but as a strategic choice? The idea, while speculative, has precedent in discussions of the Fermi Paradox. If intelligent civilisations exist elsewhere, why have we not heard from them? One proposed answer is the "zoo hypothesis," which suggests that advanced civilisations may be observing us silently, choosing not to interfere or reveal themselves—much as humans might observe endangered wildlife without intruding. Another is the "dark forest" hypothesis, inspired by Liu Cixin's *Three-Body Problem* trilogy, which posits that all life in the universe conceals itself to avoid destruction, assuming that any other civilisation is a potential threat.

In both cases, camouflage is not biological but technological—a deliberate cloaking of emissions, a refusal to transmit, a masking of activity. Just as a submarine silences its sonar to avoid detection, a spacefaring civilisation might dampen its thermodynamic or electromagnetic footprint. The consequence for our search is stark: absence of evidence becomes almost meaningless. If the most advanced beings are hiding, the louder, messier biosignatures we might detect would come only from the naïve or the newly emergent.

Yet even on a biological level, deception could be sophisticated. Imagine an alien microbe that mimics the spectral signature of inert minerals, or a photosynthetic organism that reflects light in the infrared rather than the visible spectrum. Such traits could evolve simply because they confer protection in a competitive environment. An organism that avoids triggering the predatory response of another life form—biological or machine—may be more likely to survive. We already see analogous behaviours in Earth's microbial world. Certain bacteria cloak themselves in host proteins to evade immune detection; some even alter their surface markers dynamically. If single-celled organisms can evolve such strategies on Earth, what might a billion years of alien evolution produce?

The implications of this for astrobiology are unsettling. It suggests that the silence we perceive may not reflect emptiness but ingenuity. That we might already have encountered alien life—through anomalous data, strange isotopic ratios, or unexplained atmospheric chemistry—and dismissed it because it did not meet our expectations. In this sense, our greatest obstacle may not be technical but cognitive: the inability to recognise unfamiliar forms of life because we are too committed to familiar categories.

The ethical ramifications of this possibility are equally profound. If alien life is hiding, either biologically or technologically, what responsibility do we bear in seeking to expose it? Does our right to discover extend to beings that might prefer anonymity? Could our methods of exploration—high-powered radar, spectroscopy, planetary landers—be experienced by others as intrusive or even hostile? This may seem fanciful, yet it touches on a broader ethical principle: that observation is never neutral. To look is to act. To detect is to disrupt. In a universe that may be inhabited, even modestly, our instruments are not merely passive—they are emissaries of intent.

This raises uncomfortable parallels with Earth's colonial history. The arrival of explorers often brought not understanding, but disease, exploitation, and cultural erasure. If we are not careful, the search for life beyond Earth could replicate this pattern—assuming that all beings are discoverable, knowable, and eager to be known. The possibility that life is hiding challenges not just our scientific methods, but our ethical posture. It asks whether we are prepared to accept silence not as an absence but as a boundary. Whether we can honour alien autonomy, even when it frustrates our curiosity.

It also forces us to consider the inverse: might we be the ones hiding? As our own technological civilisation matures, debates about broadcasting our presence—so-called Messaging to Extraterrestrial Intelligence (METI)—have become increasingly contentious. Some argue that we should signal our existence boldly; others fear that doing so could invite existential risk. Implicit in this debate is the recognition that intelligence might include a survival strategy of silence. That to be advanced is not necessarily to be loud.

In the end, the notion of life that lies—through mimicry, camouflage, or intentional deception—is not merely an edge-case hypothesis. It is a mirror. It reflects the limits of our perception, the assumptions embedded in our tools, and the ethics that govern our inquiry. It suggests that the cosmos may be richer and more reticent than we imagined, populated not only by beacons but by whispers, not only by sentience but by secrecy. To find what hides, we must not only look harder. We must learn to see differently.

This demands a new kind of scientific humility. One that recognises that life may not want to be seen. That the unknown may remain so not because it is absent, but because it is protected—by choice, by strategy, or by nature itself. In facing that possibility, we are challenged not to retreat from the search, but to refine it—to cultivate not only more sensitive instruments, but more imaginative minds, and more respectful hearts. For the silence of the stars may yet conceal not emptiness, but presence. Not denial, but discretion. And in that silence, life may wait—not to be found, but to be understood.

Toward a New Framework for Detection

The quest to detect life beyond Earth has long been guided by a deceptively simple idea: if life is out there, it should leave behind signs—biosignatures—in the atmospheres, surfaces, or electromagnetic emissions of other worlds. These signs, often framed in terms of familiar molecules like oxygen, methane, or chlorophyll analogues, serve as proxies for living processes. Yet as our understanding of both biology and planetary systems deepens, so too must our approach. The traditional model—searching for individual chemical species in atmospheric spectra—is increasingly recognised as narrow, Earth-centric, and vulnerable to misinterpretation. If we are to confront the possibility of alien life in its full, alien complexity, we must adopt a broader, more integrated framework for detection—one that draws from systems theory, information science, and thermodynamic principles like entropy. Only then can we begin to recognise life not just as a set of molecules, but as a dynamic process embedded within its environment.

The limitation of our current biosignature paradigm is not a lack of ingenuity, but a lack of breadth. Most detection strategies assume that life elsewhere will resemble life here—carbon-based, water-solvent, and reliant on metabolic processes that produce familiar byproducts. This assumption has yielded practical benefits: it allows scientists to target specific gases and design instruments accordingly. But it also constrains our field of view. It assumes that life's signature must be chemically conspicuous, that it must be loud rather than subtle, and that it must follow biochemistry we already understand. It is not unreasonable to imagine that these assumptions could blind us to the presence of life unlike anything we have seen.

The emerging solution is not to abandon chemical biosignatures altogether, but to embed them within a much richer conceptual matrix. Systems theory offers a powerful framework for doing just that. Rather than focusing on isolated variables—such as the presence or absence of a particular molecule—systems theory encourages us to consider the relationships between components. It is concerned with feedback loops, network stability, homeostasis, and emergent behaviours. Life, in this view, is not a substance but a process: a set of dynamically regulated interactions that maintain themselves far from thermodynamic equilibrium. A planet that sustains such a system might not emit obvious biosignatures, but it could exhibit patterns of regulation that resist purely abiotic explanation.

Earth itself provides a compelling case study. James Lovelock's Gaia hypothesis, though controversial in its teleological overtones, was fundamentally a systems-theoretical insight. Lovelock noted that Earth's atmosphere is in a striking state of chemical disequilibrium. The simultaneous presence of methane and oxygen, for instance, suggests active replenishment. More than that, the Earth system appears to modulate its climate, composition, and surface conditions in ways that have preserved habitability over billions of years. This regulation is not the result of any central planning, but of decentralised biological processes interacting with physical systems. Applying this lens to exoplanets means asking not simply, "Is there methane?" but "Does the system as a whole behave in a way that implies internal regulation?" That is a far more profound, and potentially revealing, question.

Alongside systems theory, information theory offers another powerful tool for reimagining detection. Life is not only a thermodynamic anomaly; it is an informational structure. Living systems capture, process, and transmit information—whether through genetic codes, neural activity, or biochemical signalling. This capacity distinguishes life from random or purely entropic processes. Some researchers have proposed using measures of algorithmic complexity or Kolmogorov entropy to evaluate whether observed patterns—be they in planetary albedo, atmospheric variation, or surface textures—exhibit non-random, information-rich characteristics. A landscape shaped by biology might display statistical regularities that differ markedly from those produced by wind, water, or tectonics.

Consider the phenomenon of Turing patterns: spatial structures that arise in reaction-diffusion systems and are common in biological morphogenesis. Such patterns—seen in animal coats, coral growths, and even vegetation rings—are not easily replicated by purely abiotic processes. If we were to observe similar surface structures on another world, with the right scale and geometry, we might infer the presence of self-organising, information-processing systems. Similarly, a biosphere may modulate its environment through feedback loops that produce distinct temporal rhythms—cycles in gas concentration, surface reflectivity, or thermal emissions. These modulations, if periodic and stable, may signal the influence of an adaptive, regulating agent: life.

Entropy—the third leg of this conceptual tripod—also plays a pivotal role. Life operates by capturing energy, channelling it through ordered systems, and exporting disorder to maintain internal structure. In thermodynamic terms, this is a localised reversal of entropy increase, made possible by open-system dynamics and continuous energy input. Measuring entropy fluxes at the planetary scale could thus offer clues about life's presence. For example, a planet that absorbs starlight but re-emits it in a way that deviates from radiative equilibrium could be undergoing unusual heat transformations—perhaps as a result of biological activity.

Freeman Dyson famously proposed looking for "waste heat" as a biosignature, suggesting that advanced life might not be detectable by its light, but by its entropy signature. While originally aimed at technological life, this idea applies more broadly: life, even microbial, is an engine of entropy displacement.

These concepts—systems regulation, informational complexity, and entropy management—are not merely theoretical. They are increasingly being translated into modelling and observational strategies. The "Exo-Earth System" approach, for instance, seeks to simulate entire planetary systems using coupled models of atmosphere, geology, and hypothetical biospheres. By comparing outputs of biotic and abiotic models under similar boundary conditions, researchers can identify systemic properties—such as persistent disequilibria or anomalous stability—that correlate with life. Similarly, the development of new space telescopes such as the James Webb Space Telescope and the forthcoming LUVOIR and HabEx concepts is expanding our capacity to observe fine-grained atmospheric dynamics, reflectance spectra, and surface patterns—data which can be mined for subtle, systemic indicators of life.

However, adopting such a multidimensional framework also demands a cultural shift within science. It requires interdisciplinary collaboration—not just among astronomers, chemists, and biologists, but also with computer scientists, complexity theorists, philosophers of science, and even anthropologists. For example, the very definition of "life" is far from settled, and different disciplines bring different criteria. Should life be defined by metabolism, replication, information processing, or thermodynamic boundaries? The answer is not academic. It determines what we look for, how we model it, and how we interpret ambiguous findings. A systems-oriented, entropy-aware, information-sensitive framework does not assume a single definition but remains open to multiple operationalisations.

It also compels us to reconsider the role of falsifiability in life detection. Traditionally, science has demanded that claims be testable, and that hypotheses be capable of disproof. Yet in the search for life, especially under these expanded criteria, the line between hypothesis and heuristic becomes blurry. We may never be able to say definitively that life exists on a given exoplanet. But we may reach a threshold where life is the most parsimonious explanation for a set of systemic, thermodynamic, and informational anomalies. That shift—from yes/no detection to probabilistic inference—mirrors the broader trend in science toward data-rich, model-driven reasoning.

Such a shift carries not only epistemic implications, but existential ones. If we begin to detect signs of life not as flashes of discovery, but as slowly accumulating layers of complexity—suggestive but not definitive—our cultural narrative around life in the universe may evolve. Rather than the dramatic unveiling of a "second genesis," we may face a subtler, more unsettling realisation: that life is common, but hidden in systems too intricate for simple detection. That the universe is alive not with beacons, but with whispers—complexities that resist easy interpretation but nonetheless speak of agency, adaptation, and self-organisation.

In that light, the call for a new framework is not just a technical adjustment. It is a call to maturity—a recognition that the cosmos is not obliged to make itself legible to our current instruments or paradigms. If we wish to find life, we must think like life: not as static observers, but as adaptive, relational, and deeply integrated thinkers. We must train our instruments not only to detect, but to discern. And we must train ourselves to see not just the obvious, but the emergent.

For in the end, life may not announce itself with a molecule or a flare. It may manifest as pattern, as persistence, as improbable harmony in a chaotic universe. To hear it, we must learn to listen not just to signals, but to systems. Not just to substances, but to stories—told in entropy, in order, and in the silent language of complex worlds becoming something more than matter.

Chapter 5
AI and the Algorithms of Discovery

A telescope alone is no longer the most powerful tool in the search for life beyond Earth. Increasingly, it is the algorithm—the quietly evolving intelligence trained to see patterns invisible to the human eye—that leads the charge. Artificial intelligence (AI), and more specifically machine learning, is not merely augmenting the discovery process; it is beginning to reshape how we perceive the cosmos. From parsing vast data streams generated by telescopes like Kepler and TESS to identifying subtle, non-intuitive signals in noise-drenched spectra, AI is advancing the field of exoplanetary science at a pace that would have seemed implausible a decade ago. But with this shift comes a provocative set of questions. As we teach machines to search for life, are we also training them to think like us—and therefore risk missing what doesn't conform to our assumptions? And more challengingly still: what happens if the first confirmable detection of alien life is made not by a human, but by an autonomous system?

The recent surge in exoplanet discoveries owes much to machine learning models that can process enormous volumes of data with precision and speed. NASA's Kepler mission alone generated light curves for over 150,000 stars, each curve comprising thousands of brightness measurements over time. Embedded within this astronomical haystack are minute dips in light—transits—caused by orbiting planets. Spotting them by eye is impossible at scale, and even traditional algorithms struggle with irregularities and false positives caused by starspots, instrumental noise, or overlapping stellar systems. Neural networks, however, trained on labelled transit data, have demonstrated remarkable efficacy in distinguishing real planetary signals from imposters. One such model, developed by Google's AI research team in collaboration with astronomers, famously identified two new exoplanets in archival Kepler data—objects that had been missed by previous analyses.

This is not merely a case of automation improving efficiency. It reflects a deeper shift in epistemology. Machine learning excels at uncovering statistical regularities that defy human intuition. It does not ask what a planet "should" look like; it looks for correlation structures that align with training patterns, even if those patterns appear random or insignificant to us. In this sense, AI becomes a new kind of observer—not bound by our cognitive heuristics or biases, but capable of extracting latent information embedded in complex, high-dimensional data. And yet, therein lies the rub: machine learning systems are only as good as their training data. They inherit not just our categories, but our constraints. If we feed them datasets built on Earth-centric assumptions—of what constitutes a biosignature, of what kinds of light curves indicate planets, of what atmospheric chemistry is "plausible"—we risk constructing a search engine optimised for finding only the familiar.

This issue—often summarised as "bias in, bias out"—has become a focal concern in the use of AI across disciplines. In astrobiology, it raises existential questions. Can a machine trained on our concepts ever surprise us with something truly alien? Or does it merely accelerate our tendency to find reflections of ourselves in the stars? The risk is not trivial. A biosphere based on ammonia, silicon, or exotic solvation chemistries might produce weak or unexpected signals—variations too subtle or "illogical" for a system calibrated on oxygen-rich Earth-like atmospheres. In such a case, even the most sophisticated AI might discard the data as noise. Worse still, it might misclassify the anomaly as a technical artefact, filtering it out before a human even sees it.

To mitigate this, some researchers are turning toward unsupervised learning—where algorithms search for structure in unlabelled data, free from predefined categories. Such models can identify outliers or anomalous patterns without being told what to look for. This approach has yielded promising results in other fields, from fraud detection to genomics. Applied to astronomy, it might help uncover signals that do not fit our existing frameworks—perhaps the signatures of life that we never thought to anticipate. But even unsupervised systems are shaped by the architectures we design and the features we choose to emphasise. Imagination, it seems, is still a bottleneck.

Looking beyond pattern recognition, the next frontier in detection may lie in the development of quantum sensors—devices that exploit quantum phenomena to achieve unprecedented sensitivity. Unlike classical sensors, which measure bulk properties like intensity or pressure, quantum sensors can detect minuscule variations in gravitational fields, electromagnetic signatures, or molecular vibrations with astonishing precision. Such instruments could allow us to analyse exoplanetary atmospheres remotely, detecting not just the presence of gases, but their isotopic ratios, vibrational states, and interaction networks. Combined with AI, these tools may enable us to "sniff" the chemical breath of distant worlds, searching for thermodynamic or kinetic signatures too faint for today's spectroscopy.

But as our tools become more capable, so too do the ethical questions they raise. Imagine a scenario in which an AI system, analysing data from a new space-based observatory, identifies an anomalous spectral pattern that strongly suggests biological activity. Human researchers are unable to replicate the finding directly—the signal is at the edge of detectability—but the AI's confidence is high, based on cross-validated correlations across multiple wavelengths and environmental parameters. Who, in this case, gets to make the call? Is it science if the discovery cannot be independently verified by human intuition or understanding? Is it prudent to broadcast the finding? Or should we wait for human-confirmable proof that may never arrive?

This is not a hypothetical problem. We are approaching a point where AI systems will not only assist in scientific discovery but may become its primary agents. They will identify candidates, prioritise observations, generate hypotheses, and in some cases make autonomous decisions about follow-up missions. In such a world, the distinction between tool and collaborator begins to blur. The possibility that the first confirmation of alien life might emerge from an algorithm—without a human immediately realising it—raises profound questions about authority, accountability, and the nature of evidence itself.

It also touches on deeper philosophical questions about consciousness, agency, and the status of non-human intelligences. If a machine detects life, and can explain its reasoning only in probabilistic terms beyond human interpretability, does the discovery "count"? If it autonomously initiates a message to a detected civilisation, who bears responsibility for the contact? These are not issues of science fiction alone. They are ethical and epistemic challenges for which our current institutions are poorly prepared.

This chapter will explore these challenges in depth. It will examine how AI is transforming the way we conduct exoplanet research, revealing worlds and patterns beyond the reach of human perception. It will interrogate the biases embedded in our models and datasets, and the risks those biases pose in the search for the unfamiliar. It will survey the cutting-edge technologies—quantum sensors, high-dimensional data analysis, semi-autonomous probes—that promise to expand the scope of our inquiry. And it will consider the unsettling but increasingly plausible scenario in which machines, not humans, become the discoverers of alien life.

Ultimately, the rise of AI in astrobiology is not just a story of efficiency or speed. It is a story about epistemology—about how we know what we know, and what happens when that process no longer passes through the filter of human understanding. We are not simply building tools to explore the cosmos. We are building agents to interpret it on our behalf. Whether this leads us to clarity, confusion, or confrontation with our own limitations will depend on how we design these systems, what questions we dare to ask, and how prepared we are to accept answers that may arrive in a form we can neither predict nor fully comprehend.

Machine Learning and the Planet-Hunting Surge

The landscape of exoplanet research has undergone a radical transformation over the past two decades. From the first confirmed detection of a planet orbiting a Sun-like star in 1995 to the thousands of exoplanets catalogued today, the field has grown from speculative theory to a data-rich discipline. But this explosion of information has also created a formidable bottleneck. With missions like Kepler, TESS, and the upcoming PLATO and Roman Space Telescope generating torrents of high-dimensional data, the challenge is no longer just collecting information—it is interpreting it. In this context, artificial intelligence, and particularly machine learning, has emerged as a revolutionary force, enabling astronomers to detect, classify, and analyse exoplanets with a depth and speed that would be impossible through traditional methods. Machine learning is not just enhancing our ability to find new worlds; it is changing the very way we understand what a planet is, how it can be found, and what it might look like.

The core principle behind machine learning is that algorithms can be trained to recognise patterns in data without being explicitly programmed to do so. This is particularly powerful in exoplanet research because the signals of interest—planetary transits, radial velocity variations, or photometric anomalies—are often buried in layers of noise, artefacts, and incomplete observations. A transit, for instance, appears as a tiny dip in the brightness of a star as a planet passes in front of it. For Earth-sized planets around Sun-like stars, this dip can be less than 0.01 percent of the star's total light. Moreover, starspots, instrumental errors, and overlapping binaries can produce similar variations. Human analysts, even aided by conventional algorithms, struggle to isolate these signals at scale. Machine learning systems, by contrast, can be trained on confirmed examples to detect statistical nuances that elude human perception.

One of the most widely cited examples of this power came in 2017, when Google's AI research team partnered with astronomers to apply deep learning techniques to Kepler data. Using a convolutional neural network—a model architecture originally developed for image recognition—they re-analysed light curves that had already been screened by other methods. The result was the discovery of two new exoplanets, Kepler-80g and Kepler-90i, the latter of which completed a star system with as many known planets as our own. This wasn't merely a technical coup. It was a conceptual shift, demonstrating that machine learning could not only replicate human analysis but exceed it, revealing planets hiding in plain sight.

Since then, machine learning has been integrated into nearly every phase of exoplanet discovery. Classification algorithms are used to distinguish planetary candidates from false positives, such as eclipsing binaries or instrumental artefacts. Regression models help refine estimates of planetary radius, orbital period, and stellar properties. Unsupervised learning techniques are even being used to cluster planetary systems into taxonomic groups, potentially revealing new insights into planetary formation and migration. This is more than automation; it is discovery through abstraction, a kind of computational intuition that can surface relationships no human would have thought to look for.

One particularly promising application is in the use of generative models, such as variational autoencoders or generative adversarial networks (GANs), which can learn to model the underlying distribution of exoplanet features. These models can synthesise plausible but previously unseen transit curves, effectively generating "counterfactual" planets that stretch our conception of what is physically possible. Such simulations are not merely academic exercises—they serve as training data for detection pipelines, allowing algorithms to prepare for future discoveries that may lie outside the domain of existing knowledge.

This capability is especially critical as we move into an era of observational diversity. Kepler focused primarily on one region of the sky, observing stars similar to our Sun. TESS has broadened that scope, surveying nearly the entire sky and targeting smaller, cooler stars. The upcoming PLATO mission will further diversify our sample by focusing on bright stars in longer observing campaigns. Each of these missions produces different types of noise, sampling rates, and photometric sensitivities. Machine learning models can be adapted to these differences through transfer learning—a technique in which knowledge gained from one dataset is fine-tuned for another. This adaptability allows us to build cohesive detection frameworks that remain effective across varied observational regimes.

However, this growing reliance on machine learning also comes with significant caveats. Chief among them is the issue of interpretability. Deep learning models, especially neural networks with millions of parameters, operate as black boxes: they produce results, but their internal logic is often opaque. In a field where reproducibility and physical understanding are paramount, this poses a challenge. How do we trust a discovery when we cannot articulate the rationale behind it? How do we communicate findings to the broader scientific community—or the public—when the reasoning is statistical rather than mechanistic?

Efforts to address this have led to the development of explainable AI (XAI) tools tailored for astrophysical applications. Techniques such as saliency mapping and layer-wise relevance propagation help researchers visualise which parts of a light curve or which features of a spectrum contributed most to the model's decision. These tools not only enhance confidence in machine-made discoveries but also open new avenues for scientific insight. They can reveal which features are genuinely diagnostic of planetary transits, or whether certain periodicities in the data are being misinterpreted.

A second concern is the potential for bias, both in the data and in the models themselves. If a machine learning system is trained primarily on Earth-like transits around Sun-like stars, it may struggle to identify more exotic configurations: planets with irregular orbits, circumstellar disks, or complex light-curve signatures produced by rings or moons. This bias is not malicious; it is structural. It reflects the limitations of our current datasets and our anthropocentric expectations. To combat this, researchers are increasingly turning to simulation-based training, injecting synthetic anomalies into real data to force models to learn general principles rather than specific patterns. This approach trains algorithms to expect the unexpected—to look not only for familiar shadows, but for strange glimmers in the dark.

There is also a sociological dimension to this shift. The incorporation of machine learning into astrophysics challenges traditional hierarchies of expertise. In the past, planet hunting was the domain of skilled astronomers poring over light curves and spectral lines. Today, breakthroughs often emerge from interdisciplinary collaborations between physicists, computer scientists, and data engineers. The future of exoplanet discovery will be shaped as much by algorithmic innovation as by telescope design. This demands a new kind of literacy—scientists who are fluent in both the language of astrophysics and the grammar of machine learning. It also requires new ethical frameworks, especially as AI systems gain more autonomy in prioritising observations or making preliminary classifications.

Perhaps most provocatively, machine learning introduces a new cognitive agent into the scientific process. The algorithms do not merely extend our capabilities—they reshape our epistemic landscape. They teach us to ask different questions, to look in different places, to notice different patterns. This co-evolution of human and machine understanding is subtle but profound. It suggests that the future of exoplanet science may not be guided solely by human imagination, but by a form of intelligence that complements, and occasionally challenges, our own.

As we stand on the threshold of deeper exploration—into the atmospheres of exoplanets, the dynamics of multi-planet systems, the chemistry of potentially habitable worlds—machine learning will be our indispensable partner. It will help us navigate the deluge of data, uncover faint signals, and refine our hypotheses with unprecedented precision. But it will also demand humility. For the patterns it finds may not always align with our expectations. And the planets it reveals may not always resemble the ones we dreamed of. That is not a limitation of AI; it is its gift. It shows us the universe not as we imagine it, but as it might actually be—wilder, stranger, and more richly patterned than we ever dared to guess.

Bias In, Bias Out: Teaching AI to Imagine the Unknown

The promise of artificial intelligence in the search for extraterrestrial life is, at first glance, dazzling. Algorithms trained on vast astronomical datasets can process thousands of light curves in minutes, detect subtle spectral anomalies imperceptible to human analysts, and classify exoplanets with remarkable efficiency. But underlying this technological optimism lies a philosophical and methodological tension that cannot be overlooked: can AI systems trained on human-defined datasets and Earth-centric models ever truly help us detect something alien? Or are we inadvertently training these machines to find only what we already know—or worse, what we expect to see?

The principle of "bias in, bias out" is well understood in the broader AI community. It describes the idea that the outputs of an algorithm are fundamentally shaped by the inputs it is given and the assumptions embedded within those inputs. In fields such as hiring, policing, and healthcare, biased training data has led to discriminatory outcomes, prompting widespread scrutiny. In astrobiology, the consequences may be less overtly political but no less profound. If we feed our detection algorithms only examples of Earth-like life signatures, we risk programming our tools to ignore or misclassify truly novel forms of life—thereby replicating, at machine speed, the blind spots of our own imagination.

This concern is particularly acute in the context of supervised machine learning, where models are trained to recognise patterns based on labelled examples. In exoplanet science, these labels typically distinguish between known classes—planetary transit vs. stellar noise, habitable vs. non-habitable zone, biological vs. abiotic chemistry. But what happens when the signal does not conform to any of these categories? An atmosphere rich in ammonia, an anomalous reflectance curve, or an infrared emission spectrum with no known terrestrial analogue may simply be discarded as noise, or flagged as an anomaly without meaningful follow-up. The model has not failed in a technical sense—it has done precisely what it was trained to do. The failure is epistemic. We asked the wrong questions, so we received the wrong answers.

Consider the case of the Viking missions to Mars in the 1970s. The life-detection experiments yielded ambiguous results—some consistent with microbial metabolism, others pointing to abiotic chemistry. The scientific community, uncertain how to interpret the data, ultimately concluded that no definitive evidence of life had been found. Fast forward fifty years, and we face the same risk with AI. An algorithm might detect an unusual pattern in exoplanetary data that doesn't fit our existing taxonomies. But without a conceptual framework to interpret such anomalies, and without the humility to accept ambiguity, we may miss the significance altogether. A machine cannot transcend the epistemological boundaries we impose upon it.

This is not merely a technical limitation—it is a reflection of how science, especially astrobiology, is often guided as much by cultural assumptions as by empirical evidence. Our definitions of life tend to be carbon-based, water-dependent, and DNA-coded. These assumptions are deeply rooted in the biochemistry of Earth, which is, to date, our only sample. Yet they pervade our training data and our model architectures. We look for oxygen and methane in atmospheric spectra because these are biomarkers on Earth. We scan for chlorophyll-like pigments because they are associated with photosynthesis. We design algorithms to detect planetary sizes and temperatures that support liquid water. But what if life elsewhere is based on silicon, utilises a different solvent, or operates under principles of metabolism we do not yet understand? Our models, shaped by Earth's template, may not only fail to find such life—they may erase its presence from our datasets entirely.

Unsupervised learning offers one potential remedy. Unlike supervised models, which rely on labelled datasets, unsupervised methods seek to uncover hidden structures and correlations within unlabelled data. In the context of life detection, these models can highlight outliers—spectral signatures, thermal profiles, or transit anomalies—that deviate from known patterns. An unsupervised algorithm doesn't ask whether something is "alive" or "not alive"; it identifies that something is different. This opens a door to novelty, to signals that may not fit our categories but warrant further investigation.

Yet unsupervised learning also presents challenges. Without ground truth labels, it is difficult to validate findings. Not every outlier is meaningful—many are noise, artefacts, or statistical flukes. The burden then shifts to human interpretation, and here bias re-enters the equation. Faced with ambiguity, we often default to familiarity. Our cognitive biases lead us to discount what we cannot explain or to overfit explanations to match our expectations. Thus, even when the machine highlights the unknown, it is up to us to decide whether to explore it—or to ignore it.

One promising path forward is the integration of simulation-based training. Instead of relying solely on empirical data, researchers can create synthetic datasets representing a broader spectrum of possibilities, including hypothetical biospheres with alternative chemistries, energy systems, and ecological dynamics. These simulations serve not only to expand the range of what AI models can detect but also to stretch our own conceptual boundaries. They encourage us to imagine life not as a fixed category, but as a spectrum of adaptive strategies conditioned by local environments. By exposing algorithms to this diversity, we cultivate a kind of computational pluralism—an openness to forms of life that do not mirror our own.

This approach is already gaining traction. Researchers have begun generating simulated light curves for exoplanets with rings, moons, or atmospheres containing exotic molecules. Others are modelling hypothetical worlds orbiting M-dwarf stars, where ultraviolet radiation is intense and tidal locking may produce permanent day and night sides. These simulations are not just academic exercises; they are testbeds for rethinking life's possibilities. When fed into AI pipelines, they enable the detection of anomalies that, under a more limited framework, would be filtered out.

However, the most critical shift may be cognitive rather than technical. If we hope to teach machines to imagine the unknown, we must first learn to do so ourselves. This requires an epistemic humility—a willingness to accept that life elsewhere may not look like us, behave like us, or signal itself in ways we readily understand. It also requires philosophical courage: the ability to remain curious in the face of ambiguity, to pursue hypotheses that challenge our assumptions, and to design systems that reflect not our biases, but our questions.

Herein lies a paradox. We want AI to help us discover something fundamentally new, yet we design it to operate within the contours of what we already know. To break free from this loop, we must treat machine learning not as a crystal ball, but as a mirror. It reflects our priorities, our frameworks, and our limitations. If we hope to encounter the alien—truly encounter it—we must create systems capable of being surprised. This means cultivating algorithms that do not just categorise, but inquire. That do not just optimise, but explore. That do not merely emulate our cognition, but extend it.

To achieve this, cross-disciplinary collaboration is essential. The future of life detection will not be shaped by astronomy and computer science alone. It will require insights from biology, chemistry, philosophy of science, cognitive psychology, and even speculative design. What might a non-carbon metabolism look like? How does information processing differ in non-neuronal systems? What are the thermodynamic constraints on self-organising complexity in alien environments? These are not questions that a neural network can answer. But they are the questions that must inform how we build, train, and deploy those networks.

The problem of bias in AI is not a flaw to be eradicated, but a call to awareness. It reminds us that tools reflect their makers. As we extend our reach into the cosmos with machines at our side, we must ensure that they are not merely faster versions of our own prejudices. Instead, they should become instruments of creative unknowing—designed not only to find, but to ask. For the most important discovery may not be a distant biosphere or spectral anomaly. It may be the moment when we learn to see without assumption, to model without constraint, and to search not for ourselves in the stars—but for life in its fullest, and most unimaginable, expression.

Quantum Sensors and the Future of Detection

As we scan the cosmos for signs of life, we increasingly confront a paradox of progress: the universe offers more data than ever, but many of the signals we seek remain hidden at the limits of our instruments' sensitivity. Our current tools, magnificent though they are, often scrape only the surface of what distant worlds can tell us. In particular, the atmospheric fingerprints of potentially habitable exoplanets—the chemical whispers of oxygen, methane, water vapour, or even more exotic compounds—are often buried in noise, washed out by the glare of their host stars, or obscured by the coarse granularity of existing technologies. To push further, to move from suggestion to confirmation, we must embrace the next leap in observational precision. That leap may well come from an unlikely source: the world of quantum mechanics.

Quantum sensors, devices that exploit the principles of quantum entanglement, superposition, and coherence to achieve measurements of extraordinary accuracy, are poised to revolutionise the field of remote sensing. These instruments are not merely refinements of classical tools—they represent an entirely new class of detection, capable of accessing information that was previously out of reach. They promise not only finer resolution and greater sensitivity, but a fundamental shift in how we interact with the light and matter from distant worlds. They may allow us, quite literally, to "sniff" the atmospheres of exoplanets—molecular fragment by fragment—without ever leaving our solar system.

To understand why quantum sensors matter, it is worth briefly revisiting the constraints of current spectroscopic methods. Most of what we know about exoplanet atmospheres comes from transit spectroscopy: when a planet passes in front of its star, a tiny fraction of starlight filters through the planet's atmosphere, imprinting spectral signatures that can be analysed to infer chemical composition. This technique, however, is inherently limited. The signal is minute, and easily overwhelmed by stellar variability, instrumental noise, or interstellar absorption. Moreover, the method is biased towards large planets orbiting bright, nearby stars—worlds that may not be representative of the most biologically promising targets.

Quantum-enhanced sensors offer the possibility of breaking through these limitations. One promising avenue involves the use of quantum entanglement to improve the signal-to-noise ratio in photodetection. Traditional photon detectors rely on counting incoming particles of light, but quantum techniques can correlate photons in such a way that noise—especially that arising from thermal fluctuations or background interference—is effectively cancelled out. This enables more precise discrimination between different wavelengths, allowing scientists to resolve subtle absorption lines that would otherwise be lost. In atmospheric analysis, where distinguishing between carbon dioxide and methane might hinge on the detection of a fractional percentage difference in spectral features, this improvement could be transformative.

Another application lies in quantum interferometry, which uses entangled photons to achieve resolution beyond the diffraction limit of classical optics. By correlating the phase information of multiple light paths, researchers can construct interferometric baselines equivalent to telescope apertures far larger than physically possible. This approach is already being tested in Earth-based quantum telescopes and could, in the future, be incorporated into space-based arrays. For exoplanet study, this means the possibility of directly imaging the atmospheres of Earth-sized planets orbiting Sun-like stars—an observational feat currently beyond our reach.

Perhaps most exciting is the potential for quantum coherence-based sensing. Techniques such as electromagnetically induced transparency (EIT) and quantum weak measurement allow for the detection of molecular transitions with extraordinary specificity. Applied to exoplanetary science, this could enable the identification of trace gases at parts-per-billion levels—critical when searching for biosignatures such as nitrous oxide, methyl chloride, or dimethyl sulphide. These compounds, produced in trace amounts by biological processes on Earth, may offer stronger indications of life than more ambiguous signals like oxygen, which can also arise abiotically.

Real-world applications of quantum sensors are already emerging, though not yet in space. On Earth, quantum gravimeters are being used to map subterranean voids and track groundwater movement. Quantum magnetometers are detecting biomagnetic fields in medical diagnostics. And quantum clocks, with precisions approaching one second over the age of the universe, are redefining timekeeping. These devices all rely on the manipulation of quantum states in atoms, ions, or photons—states so delicate that their coherence must be preserved with extraordinary care. Translating this precision into space-based platforms is a significant engineering challenge, requiring robustness against radiation, temperature fluctuations, and mechanical vibration. But the trajectory is clear: quantum sensing is maturing from a laboratory novelty into a practical technology.

In the context of life detection, the implications are profound. Consider the hypothetical detection of phosphine on a distant exoplanet—an anomaly first reported in 2020 in the clouds of Venus, and one that ignited a firestorm of debate. Phosphine is a potential biosignature gas, yet its detection was marginal, its signal at the edge of instrumental sensitivity, and its interpretation highly contested. With quantum sensors, such ambiguity could be reduced. Not only could we detect the gas with higher confidence, but we could analyse its isotopic composition, its vertical distribution in the atmosphere, and its temporal variation—all key parameters for distinguishing biological from abiotic sources.

Quantum sensors also invite new observational strategies. Instead of relying solely on planetary transits or radial velocity measurements, we could employ time-resolved spectrometry and light-field reconstruction, capturing the reflected or emitted light of an exoplanet across multiple temporal and spatial dimensions. This would allow for dynamic characterisation—watching weather patterns shift, seasons change, or even surface features rotate into view. Such richness of data could offer clues not just about habitability, but about active ecological or geological processes.

Yet with this power comes a new kind of interpretive challenge. The increased resolution and sensitivity afforded by quantum sensors will undoubtedly yield data of unprecedented complexity. AI and machine learning systems will be essential to processing and interpreting this information, creating a symbiotic loop between quantum hardware and algorithmic software. But as we have seen, algorithms are only as insightful as their assumptions allow. To make full use of quantum-enhanced data, we will need models that are not only chemically and physically accurate, but epistemologically expansive—capable of accommodating life as we don't yet know it.

Moreover, there is an ethical dimension to this leap in capability. As our tools become more powerful, so too does our responsibility to interpret their outputs wisely. A premature claim of life based on marginal spectral evidence could erode public trust in science; a missed signal due to overly conservative thresholds might delay a transformative discovery. In both cases, the stakes are amplified by the power of the instrument. We must therefore develop not just better detectors, but better frameworks for evaluating and communicating uncertainty.

Ultimately, quantum sensors may prove to be not just technological upgrades, but conceptual catalysts. They force us to reimagine what it means to observe, to detect, to know. They challenge the classical boundaries between object and measurement, reminding us that at the most fundamental level, the act of detection is also an act of entanglement—between observer and observed, expectation and reality. In the search for life beyond Earth, this philosophical resonance is more than metaphor. It is a call to humility, to vigilance, and to awe.

As the era of quantum-enhanced astronomy dawns, the question is not merely whether we will detect alien atmospheres with greater precision. It is whether we are prepared to recognise what we find. The signals may come not with fireworks, but with statistical subtlety—with a spectral line slightly sharper, a vibrational mode just a shade too persistent. It may be in the quiet regularity of a gas we cannot explain, or in the shimmering dance of molecules unbound by Earthly rules. To detect such signs, we will need sensors that stretch the very limits of physics—and minds that are just as willing to be stretched.

Will the First Discoverer Be Non-Human?

It is one of the most enduring images in science fiction: a human scientist, staring wide-eyed at a monitor, decoding the first definitive evidence of alien life. The scene is rich with symbolism—intellect confronting the unknown, consciousness grasping a message from the stars. But what if that moment never comes? What if, instead, the discovery is made by a machine—an algorithm parsing data on an orbiting telescope, a neural network recognising a spectral pattern, a quantum-enhanced sensor coupled with unsupervised learning that flags a biochemical anomaly invisible to the human eye? Increasingly, as artificial intelligence (AI) becomes central to scientific observation and interpretation, the answer to a question that once seemed rhetorical—*will the first discoverer of alien life be non-human?*—demands serious contemplation.

The scenario is no longer hypothetical. Machine learning systems are already outperforming humans in detecting exoplanets, classifying stellar types, and identifying anomalies in large-scale datasets. Google's collaboration with astronomers to apply a neural network to Kepler data in 2017 revealed planets that traditional analysis had overlooked. That same year, an algorithm helped identify unusual galaxies in the Sloan Digital Sky Survey that had escaped human attention. The tools we now rely upon are not just efficient—they are perceptive in ways that we are not. They have the capacity to notice patterns that defy our cognitive biases and to correlate variables across dimensions too vast or abstract for human intuition to manage.

In astrobiology, this has profound implications. The search for life involves vast datasets: light curves from transiting planets, spectra from atmospheres, radar echoes from planetary surfaces, radio emissions from deep space. These datasets are not just voluminous—they are noisy, incomplete, and ambiguous. The signal we hope to detect—a subtle imbalance in gases, an unusual isotopic ratio, a rhythmic pulsing from a distant moon—is unlikely to scream "life" in any obvious sense. It may be a statistical anomaly, buried in terabytes of telemetry, interpretable only through probabilistic inference. In such a context, the first signal of life may very well be flagged by an autonomous system before any human knows to look.

This prospect raises an immediate epistemological challenge: what does it mean to "discover" something if the discoverer is not a conscious being? Traditionally, discovery is not merely the registration of a signal—it is the act of recognition, of understanding that something observed is both novel and significant. Yet AI systems are increasingly capable of identifying and labelling novelty without direct human input. If an algorithm correlates atmospheric methane and oxygen on an exoplanet and determines that the observed combination is statistically incompatible with known abiotic processes, is that not, in some sense, a discovery? If the system issues a confidence rating, generates a report, and even recommends follow-up observations—all without a human prompting it—hasn't it already crossed the threshold of autonomous scientific agency?

The question of credit is not merely academic. Scientific prestige, funding, and public recognition are built on the notion of intellectual authorship. Who "gets credit" for a discovery shapes who gets to define its meaning, set its agenda, and narrate its implications. If a machine is first to detect alien life, does the credit belong to its creators, its operators, its institutional context—or does the concept of credit itself require revision? Some might argue that credit should always trace back to the humans who designed the system. After all, AI does not function in a vacuum; it inherits its architecture, training data, and evaluative criteria from human hands and minds. Others, however, contend that the very premise of discovery must evolve. Just as we now accept that instruments can extend human perception, we may need to accept that intelligent systems can extend human cognition—and occasionally surpass it.

This issue becomes even more pressing when one considers the growing autonomy of observational systems. NASA's next-generation observatories, including the Roman Space Telescope and potential future missions like LUVOIR and HabEx, will generate datasets of unprecedented complexity. Onboard data triage may be necessary—AI systems deciding in real time which signals to store, which to discard, and which to elevate for further scrutiny. In such environments, the line between observer and filter blurs. What we see depends not just on what the telescope captures, but on how its embedded intelligence chooses to interpret and prioritise that data.

This new paradigm challenges the traditional scientific method, which values transparency, reproducibility, and the clear articulation of inference. Deep learning systems, while powerful, are often opaque. Their decisions are the product of high-dimensional weight matrices, not clear logical chains. If a neural network flags a biosignature based on its internal model, but we cannot reverse-engineer the rationale, do we accept its conclusion? How do we verify it? Who adjudicates the claim? These are not just technical questions—they are philosophical and institutional ones. They demand new protocols for interpretability, explainability, and scientific governance.

There is also a cultural dimension to consider. Humanity has long imagined that the first contact with extraterrestrial life would be a moment of profound existential significance—a shared event, uniting people across nations and disciplines. The idea that this moment might unfold in a data centre, with a machine quietly issuing an alert that is initially overlooked or misunderstood, feels anticlimactic, even alienating. It disrupts the narrative of human centrality, suggesting that we may be passive recipients of one of the most transformative discoveries in history. This emotional resistance is not trivial. It shapes how such a discovery would be received, interpreted, and acted upon.

At the same time, the involvement of non-human intelligences in the detection of alien life offers a compelling symmetry. It would be a case of life—not necessarily conscious, but capable—recognising life. A mirror held up not to human minds alone, but to the cognitive extensions we have created. In that moment, the definition of intelligence, perception, and even communication could shift. Perhaps what matters is not who or what detects life, but how we respond to the detection—how we integrate it into our worldview, how we approach the question of interspecies ethics, and how we reflect on the tools and minds that made it possible.

We are not the first to grapple with such questions. The SETI community has long debated the role of automation in signal detection. The "Wow! signal" of 1977—a strong, narrowband radio signal detected by a radio telescope in Ohio—was flagged by an algorithm, but only gained fame because a human astronomer noticed it on a printout and scrawled "Wow!" in the margin. That single moment of human attention became iconic. But in the future, the equivalent may be buried in a digital stream of alerts, noticed only retrospectively—if at all.

What we must prepare for, then, is not just a new kind of discovery, but a new way of discovering. One in which agency is distributed across systems, in which credit is collective or diffuse, and in which the distinction between observer and instrument is increasingly porous. This requires rethinking our institutional structures, our epistemic norms, and our philosophical assumptions. It also demands humility. For if the first recogniser of alien life is not a person, but a machine—trained by us, but no longer bound by our limitations—it will mean that our search has outgrown the shape of our expectations.

And perhaps that is fitting. For if life elsewhere is unlike anything we've imagined, it stands to reason that its discoverer might also be unlike anything we've prepared for. Whether a telescope, an algorithm, or an emergent synthesis of both, that first encounter will still reflect something essential about who we are—not in our uniqueness, but in our reach. We may not be the first to see. But we will be the ones to understand what the seeing means. And in that, the human role endures—not as the sole discoverer, but as the one who asks, reflects, and responds.

The Ethics of Autonomous Discovery

As we move deeper into the age of machine intelligence, our most profound scientific frontiers are increasingly being shaped not by human hands, but by systems designed to extend—or even supersede—our perceptual and cognitive limits. Nowhere is this transformation more acute than in the search for extraterrestrial life. The rise of autonomous systems in observational astronomy, data interpretation, and mission control opens up exhilarating possibilities. But it also prompts an urgent ethical reckoning. What happens when machines interpret signals we cannot independently verify? What responsibilities do we bear when the first tentative contact with alien intelligence might not pass through human cognition at all? What does it mean—morally, philosophically, and procedurally—if an AI system not only detects the first hint of life beyond Earth, but also decides how we respond?

The integration of artificial intelligence into astrobiology is not speculative; it is already underway. Systems powered by machine learning are used to sift through massive datasets from missions like Kepler, TESS, and the James Webb Space Telescope, flagging anomalies in light curves and spectral lines far more efficiently than human analysts. These models are often trained on millions of examples, allowing them to generalise and extrapolate in ways that are opaque to their creators. In many cases, scientists cannot fully explain why a given algorithm has classified a data point as significant—they simply verify, after the fact, that it was correct. This gap between output and interpretability—what philosophers and engineers alike refer to as the "black box" problem—becomes ethically charged when extended to the possibility of alien contact.

Imagine, for instance, that an autonomous system equipped with deep learning capabilities analyses the atmospheric data from a distant exoplanet and detects a chemical signature indicative of metabolic activity. The system assigns a high probability to the presence of life, generates a report, and recommends follow-up observations. However, the signal lies at the very edge of detectability—too faint for human scientists to independently replicate with current instrumentation. Do we act on the machine's analysis? Do we classify the finding as tentative, probable, or confirmed? And critically, who decides? These are not merely operational questions—they are questions of epistemic legitimacy. At what point does a machine's "belief" constitute sufficient grounds for human action?

Now extend this scenario further. Suppose the system not only detects the signal but is configured to respond autonomously—to send an acknowledgement, to initiate a standardised broadcast, or to alter its observational protocol to prioritise ongoing interaction. This is not science fiction. AI agents are already being used to manage complex communications infrastructure, from satellite relays to interplanetary probes. It is conceivable that, in the not-so-distant future, a sufficiently autonomous scientific platform might interpret a detected signal as intentional and respond in kind—without prior human approval. This possibility transforms the ethics of machine agency from a question of interpretive trust to one of diplomatic authority.

The ethical questions this raises are manifold. Who bears responsibility for a machine's decision to engage? What constitutes informed consent when the "actor" is an artificial entity operating on pre-programmed directives? And perhaps most fundamentally: can an algorithm speak for humanity?

Historically, scientific discovery has carried with it implicit frameworks of authorship, accountability, and deliberation. When

Carl Sagan and colleagues composed the Arecibo message in 1974, every aspect of its content—numerical patterns, molecular formulas, representations of DNA—was carefully curated through human consensus. The act of messaging extraterrestrial intelligence (METI) was considered a profound expression of collective identity. By contrast, a machine-generated message may be optimised for efficiency or clarity, but stripped of cultural nuance, ethical reflection, or symbolic meaning. An AI might decide that a binary pulse conveying a string of prime numbers is the most logical format for contact—but would such a signal carry our essence, or merely our logic?

There is a risk, then, of reducing one of humanity's most significant moments—the recognition or initiation of interstellar contact—to an automated process, devoid of the deliberative, ethical, and even emotional dimensions that define human interaction. This is not merely an aesthetic loss. It has real implications for how that contact is perceived by others, whether human or alien. An AI-initiated response might be misinterpreted as aggressive, indifferent, or opaque. It might establish a precedent or a protocol that future humans cannot easily revise. In diplomacy, tone is everything. And yet machines have no tone—only function.

The debate over autonomous contact also intersects with broader concerns about transparency and control. Algorithms that learn from vast datasets can develop internal representations and decision pathways that are incomprehensible even to their developers. This creates a crisis of interpretability. If we cannot explain why a system has identified a given signal as biologically significant, or why it has prioritised one communication strategy over another, how can we justify acting on its conclusions? The principle of "explainability" is a cornerstone of scientific legitimacy and democratic accountability. Without it, we risk entering a world in which decisions of civilisational importance are made without meaningful oversight.

To be clear, this is not an argument against autonomy per se. In many contexts—such as remote planetary missions where communication delays make real-time human intervention impossible—autonomy is a practical necessity. But autonomy without ethics is automation, not agency. We must embed within our systems not only the capacity to act, but the discernment to pause, to seek confirmation, and to elevate ambiguity rather than obscure it. This requires interdisciplinary collaboration: between engineers, ethicists, scientists, and philosophers. It requires institutional foresight, legal frameworks, and, perhaps, international treaties. The protocols for AI-augmented discovery and response must be as robust as the technologies themselves.

There are already calls within the scientific community for more stringent governance of AI systems involved in space exploration. Some proposals suggest that no machine should be authorised to transmit interstellar messages without prior human review. Others advocate for layered autonomy—where machines can initiate internal analyses and flag priorities, but require human authorisation to engage in external communication. These safeguards are not about limiting progress; they are about preserving meaning. The first contact with alien life, even if initiated by a machine, should be comprehensible and intentional from the standpoint of the human community it purports to represent.

We must also resist the temptation to conflate automation with objectivity. A machine's analysis is not free from bias—it is shaped by training data, model architecture, and embedded assumptions. A neural network trained on Earth-centric models of life may overlook non-terrestrial signatures. A communication protocol optimised for data compression may erase the very ambiguity that makes a message meaningful. Ethical discovery demands more than signal detection; it demands the capacity to recognise when a signal challenges our framework, and to respond not with certainty, but with curiosity and care.

In the final analysis, the ethics of autonomous discovery compel us to ask not only what we seek, but how we wish to find it—and how we will be seen when we do. Machines may become our eyes and ears in the cosmos, but they must not replace our voice, our values, or our vigilance. As we delegate more of our exploratory power to artificial agents, we must not abdicate our responsibility to interpret, to reflect, and to decide. For contact, whether biological or intelligent, is not just a scientific milestone. It is a moral encounter. And the ethics of that encounter must begin here, on Earth, with the decisions we make about the machines we send to speak on our behalf.

Chapter 6
The Cosmic Mirror: Psychology and the Need to Know

Long before we developed the scientific means to detect distant exoplanets, we looked up at the night sky and imagined we were not alone. The urge to populate the heavens with other minds—whether godlike, monstrous, or companionable—has echoed through myths, religions, and speculative fiction for millennia. What modern science has done is sharpen the lens through which that impulse is expressed. With telescopes orbiting Earth and algorithms scouring data for biosignatures, the search for alien life has become an empirical endeavour. Yet beneath the surface, our fascination remains deeply psychological. This chapter delves into that inner terrain—not of space, but of self—exploring how the quest for alien intelligence serves as a mirror for human longing, fear, and identity.

At the heart of this emotional architecture lies what some scholars call the *loneliness hypothesis*: the idea that our drive to discover extraterrestrial life is not just curiosity cloaked in scientific garb, but a response to existential solitude. The universe, as far as we know, is incomprehensibly vast and, to date, stubbornly silent. For a species that evolved through social bonding, that silence registers as more than a data point—it becomes a psychological ache. The question "Are we alone?" is as much about our need for connection as it is about astronomical probability. Discovering life elsewhere would not merely expand our understanding of biology; it would confirm that the cosmos is, in some way, shared. And if that life were intelligent, it might suggest that mind—the very thing that makes us feel unique and burdened by awareness—is not a singular anomaly, but a cosmic tendency.

This desire to connect, however, often comes filtered through the lens of projection. In science fiction and even in speculative science, aliens frequently reflect us back to ourselves. They are militaristic or peaceful, hierarchical or collectivist, rational or spiritual—in ways that map suspiciously well onto our own political dichotomies or cultural anxieties. This anthropomorphism is not simply a failure of imagination; it is a psychological mechanism. Faced with the incomprehensibly alien, we instinctively impose familiar patterns. We imagine bipedal forms, emotional expressions, even language and morality. Carl Sagan's depiction of benevolent intelligences in *Contact*, the brutal hive minds of *Starship Troopers*, or the elegant but dangerous Vulcans and Klingons of *Star Trek*—these are less about plausible xenobiology and more about holding a mirror to ourselves. The alien becomes a narrative device through which we explore what it means to be human.

In this context, the Fermi Paradox takes on an added dimension. The question—*If extraterrestrial life is probable, why haven't we found any evidence?*—is not only a scientific riddle. It resonates like a test of cosmic self-worth. Are we being avoided? Overlooked? Have other civilisations come and gone, their signals swallowed by time and entropy? Or worse—are we alone because something intrinsic to intelligence leads inevitably to self-destruction? Each speculative answer to the paradox carries an emotional weight. Optimistic explanations suggest we're early in the cosmic timeline, that great contact awaits us. Pessimistic ones hint at filters we have yet to pass—or may not survive. In all cases, the paradox becomes a psychological crucible, forcing us to confront our fragility, our significance, and our responsibility.

The psychological stakes rise further when we consider how the presence—or absence—of alien life shapes our worldview. Hope and dread interweave. If intelligent life is common, it could signal a universe teeming with possibility, where consciousness is not an aberration but a thread in the cosmic fabric. Such a discovery could elevate our sense of purpose and belonging, even reframing human history as one expression among many of a broader narrative of sentient evolution. Conversely, the realisation that we are alone might sharpen the urgency of our self-preservation. If Earth harbours the only known life, then we become its stewards by default, and extinction—ours or any species'—takes on a profound finality.

There is also the subtler possibility that discovering alien life might unsettle rather than reassure. Imagine a civilisation millions of years more advanced, with forms of cognition, ethics, or technology so beyond our comprehension as to render meaningful dialogue impossible. Would that expand our identity or annihilate it? Contact might induce not just awe, but cognitive dissonance—a breakdown in our frameworks for understanding intelligence, morality, even reality itself. This is the darker side of the contact narrative: not the fear of physical domination, but the existential threat of irrelevance.

Throughout all of this, aliens serve as the ultimate "Other"—the abstract representation of what lies outside the self, the tribe, the species. As such, they function psychologically like the strangers in our myths: objects of fascination and terror, projection and deflection. They are whatever we most wish for and most fear. They are utopians who offer enlightenment, or invaders who expose our weakness. They are teachers, saviours, adversaries, or experiments gone wrong. In each version, they symbolise the boundary of our conceptual universe—and the challenge of transcending it.

This chapter will unpack these themes with the tools of psychology, philosophy, and cultural analysis. We will examine how the search for alien life serves as a proxy for deeper emotional needs: the need for recognition, for narrative, for connection. We will explore how our metaphors shape our hypotheses, and how scientific ambition is entangled with psychological projection. We will ask whether the silence of the stars is a mystery to be solved—or a mirror held up to our own unacknowledged questions.

In doing so, we will come to see that the search for alien life is not merely an outward expedition. It is an inner inquiry. The alien is not only out there, but within: a symbol of the unknown parts of ourselves, a placeholder for questions we have yet to formulate. Whether or not we find other minds among the stars, the act of looking reshapes who we are. It confronts us with our hopes, our fears, and the profound need not just to know—but to be known.

The Loneliness Hypothesis

The night sky, vast and unrelenting in its silence, evokes a sense of awe that borders on the existential. Beneath its infinite canopy, humanity has long asked a deceptively simple question: are we alone? That question, scientific on its surface, is profoundly emotional at its core. It is less a technical inquiry into the distribution of life across the cosmos than a psychological mirror reflecting our own desire for connection. This is the essence of what some scholars and philosophers refer to as *the loneliness hypothesis*—the idea that our fascination with alien life is not merely about expanding biological understanding, but about resolving a deeper fear: the fear of cosmic solitude.

The loneliness hypothesis frames the search for extraterrestrial life as an antidote to existential isolation. In a universe that appears, thus far, silent and indifferent, the discovery of another sentient species—another mind—would be more than a scientific breakthrough. It would be a profound metaphysical relief. It would confirm that consciousness, intelligence, and perhaps even empathy are not anomalies confined to one pale blue dot, but threads woven through the wider fabric of reality. It would suggest that our most human qualities are not freak occurrences, but echoes of a larger pattern.

This emotional undercurrent explains why so much popular discourse surrounding alien life is charged with sentiment. We speak not just of discovery, but of contact, as if what we yearn for is not merely to know, but to be known. The notion of a "galactic community," so pervasive in science fiction, reflects not just imaginative optimism but psychological compensation. If we can imagine a universe where intelligence arises in many places, we can imagine a future in which humanity is not the apex but a participant, not a lonely exception but a welcome member of a greater whole. It offers hope that our struggles—our wars, our ecological crises, our moral failures—are not the final word, but the growing pains of a civilisation on the brink of a broader kinship.

This longing for cosmic companionship is evident not only in fiction but in the institutional frameworks of science. The Search for Extraterrestrial Intelligence (SETI), founded in the 1960s, has always been as much about aspiration as methodology. The very act of listening—of pointing our radio telescopes at the stars and waiting for a signal—carries the symbolic weight of faith. Not religious faith, perhaps, but a secular version of it: the conviction that we are not speaking into a void. The iconic Golden Record launched aboard the Voyager spacecraft, containing music, images, and greetings from Earth, is not a scientifically rigorous experiment. It is a message in a bottle, flung into the cosmic ocean in the hope that someone—*anyone*—might be there to hear it.

Yet the loneliness hypothesis also reveals the fragility of that hope. For every utopian vision of contact, there exists a darker psychological counterpart: the fear that we are indeed alone. This fear is not rooted in egotism, but in its opposite. It arises from the suspicion that humanity's intelligence is not a sign of cosmic purpose, but an accidental flare against an otherwise barren backdrop. If no one is out there—no other minds to share in the experience of being, no confirmation that we are part of a wider cognitive ecology—then the weight of meaning rests entirely on our shoulders. The silence of the cosmos becomes not just an unanswered question, but a burden.

This existential weight can be seen in how we interpret ambiguous data. Consider the intense excitement—and subsequent disillusionment—that followed the discovery of the Wow! signal in 1977, a strong narrowband radio signal detected by the Big Ear radio telescope. Its origin remains unexplained, but no confirmation ever followed. Similarly, reports of methane on Mars, possible microbial fossils in Martian meteorites, or phosphine in Venus's atmosphere are met with an outsized emotional response. Each possibility carries the promise of validation, and each inconclusive outcome reinforces a nagging sense of absence.

What makes this emotional dynamic especially poignant is that it is not purely reactive; it is anticipatory. We project onto the unknown not only our questions but our needs. The vastness of space becomes a canvas for our internal narratives: of abandonment, of redemption, of longing for contact that transcends borders, species, even biology. The psychological significance of these projections cannot be overstated. They reveal that what we are searching for is not only evidence of life, but reassurance that life itself matters beyond the local confines of our own planet.

This emotional entanglement has consequences for how we conduct the search. It affects how we allocate resources, how we interpret data, and how we frame public discourse around space exploration. The Mars missions, for instance, have always been freighted with symbolic importance disproportionate to their scientific output. Each rover that traverses the red dust carries not just instruments, but the weight of anticipation—perhaps this time, we will find a fossil, a microbe, some whisper that we are not the only locus of biology in the solar system. It is no coincidence that headlines often blur the line between discovery and desire, prematurely declaring that "life has been found" when the data is, at best, ambiguous.

The loneliness hypothesis also invites a more sobering reflection: that our desire for contact may, in some ways, obscure our readiness for it. What would it mean, psychologically and socially, to confirm the existence of alien life? Would it truly resolve our sense of isolation—or simply shift it to a new level? If the other is radically different—non-biological, non-communicative, or indifferent—will the discovery alleviate loneliness or compound it? Are we seeking connection, or are we seeking a mirror?

This tension is explored with nuance in philosophical and literary works. In Stanisław Lem's *Solaris*, the alien intelligence is so utterly foreign that it resists all human understanding, ultimately revealing more about the protagonists than the planet itself. Similarly, in Arthur C. Clarke's *Childhood's End*, the arrival of superior extraterrestrials brings both enlightenment and obsolescence. These narratives highlight a crucial point: discovering life elsewhere may not confirm our significance, but challenge it. The antidote to loneliness might not be comfort, but confrontation.

And yet, the impulse to seek persists. It endures not because we are certain of what we'll find, but because the act of searching affirms something essential about who we are. It is a gesture of defiance against nihilism, a declaration that meaning can be pursued even in the face of silence. Whether we are ultimately alone or not, the loneliness hypothesis reminds us that the human spirit is defined not solely by what it knows, but by what it dares to ask.

In the chapters that follow, we will explore how this yearning shapes our theories, our technologies, and our cultural imaginations. But here, in contemplating the loneliness hypothesis, we begin with the emotional core. We recognise that the search for alien life is not simply a quest for knowledge, but an expression of something deeper: the need for kinship in a cosmos that so often appears impersonal. Whether or not another voice ever answers, the call itself is a profound act of self-definition—a signal sent not only across light-years, but into the depths of our collective psyche.

Projection and Anthropomorphism in Alien Narratives

When we imagine alien life, we are not simply looking outward—we are projecting inward. Whether consciously or not, our visions of extraterrestrials have long served as psychological canvases upon which human culture, fear, and aspiration are cast. From the earliest speculative fiction to the most rigorous scientific hypotheses, depictions of alien beings are often shaped not by what we know about the possibilities of life beyond Earth, but by what we assume about ourselves. This tendency—anthropomorphism—remains a potent and often unexamined force in our cosmic imagination, influencing not only entertainment but the theoretical contours of astrobiology and SETI (Search for Extraterrestrial Intelligence). Understanding how and why we humanise the alien is essential to recognising the limits of our imagination—and to transcending them.

Anthropomorphism, in its simplest form, is the attribution of human traits, emotions, or intentions to non-human entities. It is a cognitive shortcut deeply embedded in our evolutionary psychology. Early humans who assumed agency behind rustling bushes—whether it was a predator or merely wind—were more likely to survive. That same instinct now finds expression in how we respond to animals, machines, and, pertinently, imagined alien life. The unknown is made knowable through the familiar; the radically different is rendered comprehensible through metaphor.

This psychological reflex is evident across centuries of literature and myth. Medieval accounts of life on other worlds, such as those imagined by Johannes Kepler or Giordano Bruno, frequently depicted humanoid creatures with recognisably moral or philosophical traits. In the Enlightenment era, alien intelligences were often envisioned as rational utopians—idealised reflections of human reason. By the time H. G. Wells introduced the Martians in *The War of the Worlds* (1898), the alien became not just a projection of idealism, but also of threat—technologically superior, morally ambiguous, and emotionally cold, yet unmistakably bearing the stamp of human anxieties about colonialism, scientific hubris, and societal fragility.

In the twentieth century, science fiction exploded into a medium through which cultural and psychological tensions were played out via the alien form. The Cold War, for instance, gave rise to countless narratives of invasion and infiltration, with extraterrestrials standing in for ideological enemies. In *Invasion of the Body Snatchers* (1956), the alien menace is not monstrous in appearance but invisible—embedded within familiar forms, indistinguishable from us, reflecting fears of conformity, subversion, and loss of individual autonomy. Conversely, in *E.T.* (1982), Steven Spielberg's eponymous alien is childlike, emotionally expressive, and ultimately benign—a vessel for exploring empathy, friendship, and the loneliness of otherness.

These representations are not simply flights of artistic fancy. They reveal deep-rooted tendencies to impose human moral frameworks, cognitive structures, and social dynamics onto the unknown. Even in speculative science, we find similar patterns. The famous Drake Equation, while probabilistic in nature, implicitly assumes a trajectory of development that mirrors human civilisation: from primitive life to technological intelligence, from radio communication to interstellar curiosity. SETI's focus on narrow-band radio signals reflects not just technical pragmatism, but a belief—largely unexamined—that alien intelligence would evolve in ways sufficiently analogous to our own to render communication plausible. We search for what we know how to find.

This assumption is not without merit. It is reasonable, even necessary, to begin a search with models grounded in the only known example of life—ourselves. But the danger lies in mistaking these models for universals. Just as early European explorers often interpreted new cultures through their own ethnocentric frameworks, risking misunderstanding or violence, so too do we risk misrecognising or overlooking alien life that does not fit our templates. A truly alien intelligence may not use symbols, tools, or even language as we understand it. Its perception of time, space, identity, or purpose might lie outside the bounds of human conceptualisation. Anthropomorphism, while comforting, narrows the field of possibility.

The psychological mechanics of projection are further complicated by our emotional relationship to the alien. Aliens are often depicted as embodying our greatest hopes or our deepest fears. They are saviours or destroyers, messengers or monsters. The Greys of UFO folklore, with their oversized eyes and telepathic communication, are often interpreted as representations of both advanced evolution and dehumanisation—intellect without empathy. Meanwhile, the utopian aliens of *Star Trek* frequently function as philosophical interlocutors, pushing humanity to examine its own values through contrast. In each case, the alien becomes a dramatic foil for the self.

Even the language of astrobiology is not immune to these tendencies. When scientists refer to "habitable zones," they often mean planets with Earth-like conditions: liquid water, a certain range of temperatures, an atmosphere with oxygen or carbon dioxide. These are not universal criteria for life—they are Earth-centric proxies. Silicon-based life, ammonia solvents, or exotic metabolic pathways may exist, but they are rarely central to detection strategies or funding priorities. Our tools, and the algorithms behind them, are calibrated to find life as we know it—reinforcing the anthropomorphic loop.

Moreover, the very notion of intelligence is deeply anthropocentric. Intelligence is often measured by criteria such as tool use, abstract reasoning, or linguistic capability—traits that map closely onto human cognitive evolution. But what if a civilisation evolved with a collective hive-mind structure, or with non-symbolic modes of thought? What if its "intelligence" manifested through biological self-regulation, or ecological equilibrium rather than technological artefacts? By assuming that intelligence must express itself through technology, especially radio waves or megastructures, we may be blind to forms of life that are intelligent in ways we have not yet imagined.

Some thinkers have attempted to counter this tendency. The British biologist Jack Cohen and mathematician Ian Stewart, in their provocative work *Evolving the Alien*, argue that our imaginative frameworks for life are far too narrow. They advocate for "xenoscience"—the systematic effort to model life that is not merely exotic but structurally alien. Similarly, astrobiologist Carol Cleland has argued for the development of a broader theory of life, one not confined to terrestrial definitions. Without such conceptual expansion, she warns, we may overlook biospheres that exist under our noses—or across light-years—simply because we lack the intellectual tools to recognise them.

Projection and anthropomorphism are not inherently flawed impulses. They are, in many ways, necessary steps in making the unfamiliar approachable. But when left unchecked, they constrain the scope of our inquiry. They lull us into searching for mirrors rather than mysteries. And in doing so, they risk turning the most expansive scientific endeavour—understanding life in the universe—into an exercise in self-reflection.

To guard against this, we must cultivate intellectual humility and imaginative elasticity. We must be willing to entertain models of life, mind, and civilisation that do not resemble our own. This means rethinking what counts as communication, intelligence, even existence. It means designing detection protocols that include non-equilibrium chemistry, unconventional energy gradients, and anomalous patterns of information flow. It means accepting that the first evidence of alien life may not be a message or a signal, but a puzzle—one that resists decoding not because it is silent, but because it speaks in a grammar we have yet to invent.

Ultimately, our representations of aliens tell us as much about ourselves as they do about the cosmos. They are artefacts of our cognitive architecture, our cultural myths, and our epistemic boundaries. Recognising this is not a failure of science—it is its deepening. It is the beginning of a more mature astrobiology, one that embraces not only the search for others, but the transformation of self that such a search entails. For in stepping beyond anthropomorphism, we may not only discover new forms of life—we may also learn to see, for the first time, just how strange and singular our own assumptions have been.

The Fermi Paradox as a Psychological Test

When physicist Enrico Fermi reportedly asked during a lunchtime conversation in 1950, "Where is everybody?" he could not have anticipated that this deceptively simple question would become one of the most enduring riddles in modern scientific and philosophical thought. Known now as the Fermi Paradox, the issue is straightforward in its premise yet haunting in its implications: given the vast number of stars in the galaxy, many of which host planets in the habitable zone, and the generous timescales over which life could develop, why have we not yet observed any evidence of extraterrestrial civilisations? The paradox, while rooted in astronomical statistics, has taken on a far broader cultural and emotional resonance. It has evolved into a kind of psychological mirror, reflecting humanity's anxieties about isolation, failure, and cosmic irrelevance.

At its heart, the Fermi Paradox is not just about the absence of contact—it is about the emotional significance we attach to that absence. The silence of the cosmos does not merely baffle; it wounds. It echoes like a void in the psyche, raising uncomfortable questions: Is intelligence inherently self-destructive? Have others been here and gone? Are we being ignored, or are we simply too primitive to notice the signs? These queries, while dressed in scientific garb, are laced with existential unease. The paradox does not only ask us to locate the aliens—it asks us to locate ourselves.

In this sense, the Fermi Paradox functions much like a psychological test, akin to a Rorschach inkblot. Its ambiguity allows for a wide range of interpretive projections, each revealing something about the beliefs, fears, and desires of the interpreter. For some, the absence of contact signifies human exceptionalism—a comforting idea that we are rare, perhaps even chosen. For others, it suggests a foreboding silence, a graveyard of civilisations that could not survive their own technological adolescence. Each answer, speculative as it may be, lays bare a worldview, a temperament, a moral vision.

One of the most commonly projected fears through the Fermi lens is abandonment. This is the unsettling notion that we may have arrived late to a cosmic party that has already ended—or worse, that we have been purposefully excluded. The hypothesis of the "zoo theory," for instance, posits that advanced civilisations are aware of us but choose not to make contact, observing us as we might observe an animal species in the wild. This idea can feel eerily akin to a cosmic parental neglect: we are conscious beings, yearning for dialogue, and yet we remain incommunicado. In psychoanalytic terms, this resembles the trauma of the unheard child—the one who speaks into the void and hears nothing back.

Alternatively, the paradox may be interpreted through the lens of failure—specifically, the idea that intelligence is a self-terminating phenomenon. This reading has gained traction through the concept of the Great Filter, a theoretical barrier that prevents life from progressing to the stage of interstellar civilisation. The filter could lie behind us (e.g., life is exceptionally rare) or ahead of us (e.g., technological societies inevitably destroy themselves). The latter interpretation casts the Fermi silence not as a mystery but as a warning. It suggests that the universe is not teeming with life because life has a built-in expiration date. Here, the paradox confronts us not with the fear of being overlooked, but the fear of not being able to endure. It becomes a test of our capacity to reflect, reform, and survive.

Closely related is the fear of insignificance. In a cosmos of unimaginable scale, silence can feel like indifference. Despite our scientific progress, our cultural achievements, and our aspirations, the universe offers no feedback, no confirmation that we matter beyond our own planet. This sense of cosmic apathy can be psychologically destabilising. It threatens the narrative that consciousness has intrinsic worth or that evolution towards intelligence has a deeper purpose. When faced with the Fermi silence, we are forced to contemplate the possibility that we are not just alone—we are unremarkable. That idea can be as intolerable as it is plausible.

Science fiction, which has long served as a testing ground for philosophical ideas, frequently grapples with these themes. In Arthur C. Clarke's *2001: A Space Odyssey*, alien contact is cryptic and sublime, suggesting that intelligence is part of a grand evolutionary arc. In contrast, Stanisław Lem's *Fiasco* presents contact as fundamentally impossible—so great is the gap between human and alien cognition that meaningful communication cannot occur. In both cases, the paradox becomes a crucible for human self-examination. The question is not merely whether others exist, but whether we are capable of comprehending, empathising with, or even noticing them if they do.

Real-world scientific discourse has increasingly acknowledged the psychological dimension of the paradox. SETI researchers, while grounded in radio astronomy and data analysis, often speak of the "wow factor"—the hope that one signal might transform our understanding of the universe. This hope, however, exists alongside an evolving pragmatism. Some now argue that we may need to recalibrate our expectations: that contact, if it comes at all, may be ambiguous, gradual, or asymmetrical. It may not take the form of a dramatic message, but of a strange atmospheric anomaly, a technosignature we can't quite explain, or a statistical outlier in a sea of noise. In such a scenario, the Fermi Paradox continues to test not just our intellect, but our patience, humility, and interpretive flexibility.

The paradox also has implications for how we view the future of humanity. If we interpret the silence as a sign that civilisations rarely last, then our own trajectory becomes a moral imperative. Avoiding extinction is no longer just a parochial concern—it becomes a duty to the cosmos. We may be the only flicker of intelligence within light-years, and with that rarity comes responsibility. This interpretation imbues the Fermi Paradox with a paradox of its own: it suggests that the absence of others may be what calls us most strongly to act with foresight, restraint, and purpose.

Importantly, the Fermi Paradox does not necessitate pessimism. For some thinkers, the silence is not frightening, but liberating. It means we are the authors of our own meaning, unencumbered by comparison or oversight. It means that our creative and scientific endeavours take on cosmic significance by default. If we are alone, then everything we do matters more, not less. This is a profoundly humanistic reading—one that sees the search for others not as a way to escape solitude, but as a way to deepen self-awareness.

In this light, the Fermi Paradox becomes less a mystery to be solved and more a space of reflection. It confronts us with our limitations, our aspirations, and our unresolved fears. It invites us to ask why we need to know we are not alone, and what it would mean if we were. It challenges us to examine the motives behind our search—not only what we hope to find, but what we hope to resolve within ourselves.

As the tools of detection become more sophisticated, and as missions like James Webb and future observatories peer deeper into exoplanetary atmospheres, the silence may become more sharply defined. Whether it remains or is eventually broken, the psychological resonance of the Fermi Paradox will endure. It is the paradox of the seeker—poised between curiosity and dread, knowledge and uncertainty, looking up into a sky that is not empty but eloquently mute.

In the end, the Fermi Paradox tests not only the likelihood of alien life, but the quality of our own attention. It asks whether we can bear the weight of not knowing, whether we can remain curious in the face of ambiguity, and whether the act of searching is itself a declaration of meaning. In that sense, perhaps the greatest insight of the paradox is not astronomical at all, but human: that in asking the question, we reveal not the absence of others, but the depth of our need—for connection, for purpose, and for wonder.

Hope, Dread, and the Search for Cosmic Meaning

To gaze into the night sky is to confront a paradox of scale and sentiment: the universe is unimaginably vast, yet to the best of our knowledge, devoid of company. It is in this silence that humanity has projected some of its most profound existential longings and fears. The search for alien life—whether microbial on a Martian plain or intelligent among the stars—is not merely a scientific enterprise. It is a search suffused with psychological weight and philosophical consequence. The possible outcomes—discovery, ambiguity, or enduring silence—each carry the power to reshape our worldview, recalibrate our sense of identity, and redirect the trajectory of our collective ambitions.

At the heart of this inquiry lies a fundamental human tension between hope and dread. The hope that we are not alone—that life, and perhaps intelligence, has emerged elsewhere—is deeply tied to a yearning for cosmic companionship. It promises to dissolve the sense of isolation that haunts our species, to affirm that consciousness is not an anomaly, but a recurring feature of the cosmos. In this light, the discovery of alien life is imagined as a confirmation that we are part of a larger biological continuum, that our existence is not absurd but aligned with a broader cosmic pattern.

This hope is not abstract. It influences scientific priorities, philosophical discourse, and even the political rhetoric of space exploration. Initiatives like the Breakthrough Listen project, the Mars Sample Return mission, and the steady march of exoplanet detection through instruments like the James Webb Space Telescope are grounded not only in methodological rigour but in a quiet anticipation: that somewhere, just beyond our reach, lies the evidence that will change everything. Optimism infuses our cultural narratives as well. From Carl Sagan's "Pale Blue Dot" to recent cinematic portrayals such as *Arrival*, there is a recurring belief that contact would not only broaden our knowledge but elevate our species—prompting unity, humility, and perhaps even wisdom.

Yet intertwined with this hopeful anticipation is a powerful undercurrent of dread. For some, the discovery of alien life—particularly intelligent life—evokes a visceral fear. What if contact is not benevolent but threatening? What if a more advanced civilisation views us as irrelevant, primitive, or expendable? These concerns have animated everything from Cold War science fiction to contemporary thought experiments like the "Dark Forest" hypothesis, which posits that civilisations remain silent not because they are absent, but because visibility invites destruction. In this framework, the cosmos is not a community but a battlefield in waiting—one where silence is safety.

Even if alien life were found to be benign or incomprehensibly distant, its mere existence might destabilise human narratives. Religious doctrines that place humanity at the centre of creation, philosophies grounded in human uniqueness, and socio-political ideologies built on assumptions of cultural supremacy could be profoundly challenged. The impact would be uneven. Some individuals and communities might embrace the expansion of perspective; others might retreat into denial, revisionism, or hostility. As with any paradigm shift, the discovery of life beyond Earth would act as a cultural litmus test, revealing the elasticity—or fragility—of our worldviews.

Conversely, the continued absence of evidence—despite increasingly sophisticated instruments and expanding astronomical catalogues—carries its own weight. Persistent silence invites a different kind of dread: the fear that we are, in fact, alone. This solitude, if confirmed, would make Earth not just a home, but the sole cradle of life in the observable universe. Such a realisation could be galvanising or paralysing. On one hand, it would bestow upon humanity an extraordinary responsibility—to preserve and protect the only known instance of sentient life. On the other, it could provoke a sense of despair, a feeling that our strivings, achievements, and ethical struggles are unobserved by any larger audience or context.

Psychologically, this tension between cosmic inclusion and cosmic solitude reflects the dual nature of our need for meaning. We long for validation from beyond, even as we fear what that validation might reveal. The absence of extraterrestrial life invites introspection: if we are alone, what do we do with that knowledge? Does it render our lives meaningless, or does it place an even greater value on our capacity to think, to feel, and to act with intention? For some, the idea of a universe without other minds is terrifying. For others, it is profoundly motivating—a call to treat consciousness not as common property, but as a fragile, precious anomaly to be nurtured.

History offers parallels to this existential bifurcation. When Copernicus displaced Earth from the centre of the cosmos, the psychological reverberations were immense. Similarly, Darwin's theory of evolution repositioned humanity as one branch among many in the tree of life. Each shift forced a recalibration of human identity—one that was often met with resistance before acceptance. The discovery of alien life, or the definitive lack thereof, would constitute the next great ontological disruption. It would either elevate us into a larger family of beings or isolate us in a singular silence that demands new narratives of purpose.

Moreover, the impact of this search extends beyond metaphysics and psychology into tangible questions of policy and future planning. If we anticipate the possibility of contact, how do we prepare diplomatically, ethically, and scientifically? Who speaks for Earth? What message do we send, and who decides its content? These are not idle speculations. METI (Messaging Extraterrestrial Intelligence) efforts have already begun, and with them, vigorous debates about the wisdom and consequences of broadcasting our presence. The act of messaging—of declaring ourselves—carries with it assumptions about the kind of contact we desire and the kind of beings we expect to find.

Alternatively, if we accept the probability of long-term solitude, what implications does this have for space colonisation, resource allocation, and intergenerational ethics? Would we redouble our efforts to become a multi-planetary species, not for glory or profit, but as a moral imperative to preserve life itself? Would we treat our ecosystems, our cultural heritage, and our shared future with greater reverence, knowing that the universe holds no backup copy?

The search for cosmic meaning, then, is not a passive exercise. It exerts a gravitational pull on human aspiration. It influences how we teach science, how we fund exploration, how we conceive of ourselves in relation to each other. It is not merely about finding life elsewhere—it is about finding perspective here. Whether contact comes in the form of microbial fossils, encoded signals, or remains forever hypothetical, the search itself redefines us. It pulls our gaze upwards, and in doing so, redirects our ethical compass inward.

In the interplay of hope and dread lies the full complexity of the human condition. We are a species that dares to ask impossible questions, that builds machines to listen to stars, that dreams of answers and fears their consequences. We seek life elsewhere not just to expand our knowledge, but to locate ourselves in a cosmos that, so far, has given us no definitive reply. This silence is not empty—it is charged with meaning. And it is in our response to that silence, in the questions we continue to ask and the futures we dare to imagine, that our deepest sense of cosmic meaning will be found.

Aliens as the Ultimate "Other"

Across cultures, epochs, and epistemologies, humanity has long been fascinated by the figure of the "Other"—that which is not us, not here, not now. From mythical beasts beyond the borders of civilisation to theological demons and divine messengers, the Other has served as a repository for everything mysterious, feared, admired, or misunderstood. In the modern era, particularly in the wake of astronomical discovery and speculative fiction, the alien has come to occupy this archetypal position in our collective imagination. As the ultimate Other, the alien is both our greatest curiosity and our most profound fear—an unknowable presence that reflects our deepest assumptions about self, difference, and the limits of understanding.

The alien, as a concept, is not simply an extrapolation of biology or astrophysics. It is a philosophical mirror, inviting inquiry into what it means to be human. By imagining beings entirely removed from the evolutionary and cultural conditions that shaped us, we confront the edges of our own identity. What is intelligence without emotion? What is morality without shared history? What is life, stripped of terrestrial familiarity? In this sense, the alien is not merely a hypothetical life form. It is an epistemic challenge, a test of our capacity to recognise forms of consciousness, culture, or value that may have no analogue in our world.

One of the most persistent themes in alien narratives—both scientific and fictional—is the dialectic between familiarity and radical difference. We are drawn to the alien because it promises revelation: new technologies, new philosophies, new truths about the universe. Yet we are equally repelled by the idea that the alien might be *too* different—so fundamentally alien that it resists all interpretation. In this duality, the alien reflects not only our aspirations for transcendence, but our fear of cognitive annihilation. The Other may enlighten, but it may also overwhelm.

Philosopher Emmanuel Levinas argued that true ethical engagement arises not from understanding the Other, but from acknowledging its irreducibility. Applied to the alien context, this suggests that the greatest respect we can offer extraterrestrial intelligence is to accept that we may never fully comprehend it. This is a sobering thought. It implies that even if contact were achieved, communication might be impossible—not because of technological limits, but because of ontological incommensurability. What if alien minds do not divide the world into subject and object? What if their sense of time is cyclical or multidimensional? What if their form of "language" is biochemical, or gravitational, or entirely abstract? In confronting the alien, we may be confronting the inadequacy of our categories.

This incommensurability often finds symbolic expression in science fiction. Consider the monolith in *2001: A Space Odyssey*, an artefact of immense significance but no clear origin or intention. Or the alien intelligence in Stanisław Lem's *Solaris*, which generates human replicas based on repressed memories but never reveals its purpose. These narratives do not resolve the alien; they preserve its mystery, using it to dramatise the limits of human understanding. The alien, in these works, is not a being to be assimilated or defeated, but a presence to be reckoned with—a sublime reminder that the universe may be stranger than we are capable of knowing.

Yet even as the alien challenges our epistemic boundaries, it also reveals our anthropocentric tendencies. We often imagine aliens in our own image: humanoid forms, hierarchical societies, recognisable emotions. This projection is not merely a failure of imagination—it is a psychological defence. By rendering the alien legible, we defang its threat. We make it a rival, a partner, or an avatar for human values. In doing so, we risk flattening the richness of possible difference into a reflection of ourselves. As the philosopher Thomas Nagel famously asked, "What is it like to be a bat?"—a being whose sensory and cognitive world is so different from ours as to be inaccessible—so too must we ask, what would it *feel* like to encounter a form of life that does not share our premises about selfhood, death, or reality?

The alien also operates as a moral test. How would we treat the radically other? Would we seek to dominate, exploit, or convert? Would we extend ethical consideration to beings whose interests, needs, or vulnerabilities we cannot intuit? These questions are not merely theoretical. They echo in our own history—of colonisation, conquest, and misrecognition. The arrival of the alien, should it ever occur, may expose not only our scientific preparedness but our ethical maturity. It would challenge our capacity for restraint, reciprocity, and humility.

Even the possibility of encountering aliens invites philosophical reflection on what constitutes community. Would the discovery of alien intelligence displace our human-centric frameworks, forcing a redefinition of moral scope? Or would it solidify in-group cohesion by demarcating a clear us-versus-them boundary? These possibilities are not mutually exclusive. In some speculative scenarios, contact unites humanity under a common banner. In others, it fractures us, amplifying divisions about how to respond, interpret, or co-exist. What is clear is that the alien, as the Other, functions not only as an external stimulus but as a catalyst for internal transformation.

This dual role—as challenge and mirror—helps explain why the alien has become such a persistent figure in popular culture. From benevolent visitors like Spielberg's *E.T.* to enigmatic entities in *Arrival*, from the violent xenomorphs of *Alien* to the multi-dimensional beings of *Interstellar*, each representation is a coded reflection of the cultural moment in which it was created. During the Cold War, aliens embodied geopolitical paranoia; during the post-human turn, they expressed anxiety about identity, embodiment, and the limits of language. In every case, the alien is both symptom and metaphor—an externalisation of internal uncertainty.

Importantly, the alien as the Other also holds transformative potential. It invites us to reimagine the boundaries of empathy, to expand the circle of moral concern, and to question the givens of our own thought. In this way, it plays a philosophical role akin to that of the stranger in ethics or the trickster in mythology. It destabilises, provokes, and reorders. And in doing so, it keeps the human project honest—reminding us that the world is not built to our specifications, and that meaning is always a provisional negotiation between self and alterity.

In philosophical terms, the alien is less an object than a concept—one that resists closure. It invites us to live with the discomfort of ambiguity, the humility of partial understanding, and the exhilaration of encounter without resolution. Whether we ever find extraterrestrial life or not, the alien—as metaphor, as mystery, as mirror—will continue to haunt our imagination. It will remain the ultimate Other, calling us to expand our definitions, test our values, and face the unknown not with domination, but with wonder.

Chapter 7
Alien Contact and the Ethics of Messaging

For centuries, the idea of contact with extraterrestrial intelligence has lingered at the edges of scientific speculation and philosophical rumination, casting long shadows across our understanding of humanity's place in the cosmos. But as our technology grows more sophisticated and our reach expands beyond the confines of our own solar system, this once-hypothetical scenario is slowly transforming into a legitimate ethical dilemma. We are no longer only listening—we are beginning to contemplate speaking. This chapter grapples with the many-layered implications of Messaging Extraterrestrial Intelligence (METI), a controversial practice that seeks not just to detect alien life, but to announce our presence to it. What is at stake, however, is not just the science of communication, but the moral and existential significance of initiating a conversation whose consequences we cannot predict.

The METI debate serves as the linchpin of this chapter, framing the central question: should we shout into the void? Proponents argue that if intelligent civilisations exist, they may be listening just as we are, awaiting a signal that confirms they are not alone. To send such a message, then, is not merely an act of communication but of hope—a leap of faith into the cosmic unknown. But critics caution that to broadcast our existence without a thorough consideration of the potential risks is reckless at best and catastrophically naïve at worst. After all, history offers many cautionary tales of asymmetric encounters between civilisations, where the technologically dominant party did not always act with benevolence. To presume that aliens would be inherently peaceful may reflect more about our own aspirations than the probable realities of interstellar contact.

At the core of this debate lies another equally complex issue: if we are to send a message, who speaks for Earth? The question is not merely procedural; it is deeply philosophical. Whose values, languages, priorities, and representations should be encoded into a message meant to encapsulate humanity? Past efforts—such as the Golden Record on the Voyager spacecraft or the Arecibo message of 1974—were developed by small groups of scientists and thinkers, largely Western, largely male, and largely situated within a narrow cultural framework. These projects, though well-intentioned and imaginative, now invite critique for their lack of global inclusivity. As we contemplate more formalised efforts to communicate across the stars, the need for a democratic, ethically considered process becomes pressing. Can a planetary consensus even be reached on what to say, let alone whether to say it at all?

The assumption that alien recipients would be peaceable or at least rational adds yet another layer of moral ambiguity. Popular media often paints alien civilisations as either benevolent teachers or existential threats, mirroring our most hopeful and fearful projections. But the truth is likely more complicated. The assumption of peace, often implicit in METI initiatives, is not guaranteed. Indeed, such an assumption may be a reflection of anthropocentric optimism—an extrapolation from Enlightenment ideals that communication is inherently civilising. But this premise ignores the possibility that an advanced civilisation might view unsolicited signals as intrusive, primitive, or even provocative. In that sense, the METI debate is as much about the psychology of risk as it is about the mechanics of communication. Are we prepared for the full spectrum of outcomes, including those that unsettle our assumptions about agency, morality, and survival?

The moral terrain becomes even more treacherous when we reverse the scenario: what if *we* are the discoverers, the respondents, the ones who must interpret a signal and decide how to engage? The act of discovery, often celebrated as a triumph of intellect and curiosity, also carries with it the burden of responsibility. Who decides what is done with the information? If the alien life discovered is microbial or ecosystemic rather than intelligent, does it possess rights? Do we impose Earth-centric frameworks of ethics upon non-Earth life? Do we colonise, quarantine, preserve, or interact? These questions call for a rigorous ethics of discovery—one that encompasses the possibility that life elsewhere may be fragile, sacred, or incompatible with our ways of thinking.

There is precedent on Earth that underscores the importance of restraint. The exploitation of uncontacted tribes, the devastation of ecosystems, and the history of imperial conquest all speak to the dangers of acting without a deep ethical consideration of the other. Translated into the context of alien life, these patterns serve as dire warnings. The first contact may not occur through grand, Hollywood-style diplomacy, but through subtle, ambiguous, and possibly irreversible steps. A probe, a signal, an atmospheric intrusion—each might constitute a kind of interstellar footprint with ethical ramifications. What we do, how we interpret, and what assumptions we bring to these acts are all part of a broader moral schema that remains largely unwritten.

This chapter also explores the use of thought experiments and simulation as tools for preparing for contact. What would you say to an alien civilisation, given a single transmission? Would you emphasise our science, our art, our values, or our vulnerabilities? How do you balance honesty with caution? How do you express the fullness of human experience without overstepping into hubris? By engaging in such speculative exercises, we surface the underlying tensions and contradictions within our collective identity. What we choose to say, or not say, reflects what we believe about ourselves, what we fear to reveal, and what we hope to become.

Importantly, the ethics of messaging and contact are not only theoretical dilemmas—they are fast becoming practical concerns. Private actors such as SpaceX, and open-source METI projects, are already engaging in initiatives that bypass international oversight. In this decentralised landscape, the risk of unilateral messaging grows, raising questions about governance, transparency, and accountability. Should such acts be subject to international regulation? Do we need a planetary protocol for interstellar communication akin to environmental or humanitarian law? The urgency of these questions will only intensify as technological thresholds are crossed and public interest deepens.

Ultimately, this chapter argues that alien contact—whether initiated by us or received from others—is not merely a scientific milestone. It is an ethical and philosophical turning point that will challenge our self-understanding, political structures, and moral philosophies. To engage responsibly with the question of contact, we must broaden our frameworks beyond the technical and incorporate insights from anthropology, theology, ethics, and postcolonial theory. We must prepare not only our instruments, but our imaginations, our policies, and our principles.

Whether contact comes in our lifetime or not, the act of preparing for it forces us to ask who we are and how we wish to be perceived. Do we reach out in fear or in hope? Do we broadcast noise or poetry? Do we assume that intelligence equals empathy, or do we confront the possibility that understanding alone may not ensure goodwill? These are the questions at the heart of the ethics of messaging—and they may be among the most consequential decisions humanity ever makes.

The METI Debate: Should We Shout into the Void?

To send a message into the cosmos is an act that straddles the boundary between science and existential theatre. Messaging Extraterrestrial Intelligence (METI) is not merely a technical exercise in radio transmission—it is a philosophical gesture, a symbolic declaration of presence. It affirms our belief that communication is possible, that intelligence elsewhere might hear us, and that contact could be mutually transformative. But within this seemingly noble aspiration lies one of the most contested debates in contemporary space policy and astrobiology. Should we, as a species, announce ourselves to the unknown? Is it an invitation to enlightenment—or a reckless broadcast to powers we cannot begin to understand?

The METI debate has emerged with increasing urgency as the technical capacity to send powerful, directed interstellar messages becomes more democratised. Unlike the Cold War-era transmissions such as the Arecibo Message of 1974, which were largely symbolic and confined to narrow audiences within the scientific elite, today's METI efforts are being pursued by independent researchers and private entities, operating beyond the scrutiny of any coordinated global governance. What was once the purview of national space agencies is now accessible to universities, tech collectives, and even hobbyists. This decentralisation has intensified the stakes, for it suggests that decisions about broadcasting humanity's existence may be made without comprehensive ethical deliberation or international consent.

Proponents of METI argue from a position of optimism and intellectual curiosity. If intelligent life exists elsewhere in the galaxy—and many astrophysicists now contend that, statistically, it almost certainly does—then contact may already be overdue. Passive listening, the current focus of the Search for Extraterrestrial Intelligence (SETI), presumes that other civilisations are actively sending. But if all are listening and none are speaking, the cosmos remains silent by default. METI advocates suggest that initiating contact may break this stalemate and signal our readiness to participate in a broader cosmic dialogue. They view messaging as a moral obligation—to extend the hand of communication not as conquerors, but as fellow participants in a universe potentially rich with life.

There is also a scientific rationale behind METI. Contact with an extraterrestrial intelligence, even through the faintest of signals, would constitute the most significant discovery in human history. The knowledge gleaned from such a communication—about physics, biology, ethics, or consciousness—could revolutionise every discipline. Advocates liken the act of sending a message to the launch of the Voyager spacecraft's Golden Record: an assertion that the pursuit of understanding outweighs the comfort of silence. To METI supporters, to remain mute is to forsake one of our most noble capacities—the ability to reach beyond ourselves in search of meaning.

However, the critics of METI are equally impassioned—and their concerns are rooted in a combination of historical precedent, evolutionary caution, and philosophical prudence. The primary objection is elegantly simple: broadcasting our presence to unknown intelligences is inherently risky. We cannot presume the intentions or moral frameworks of those who might receive our signals. History on Earth offers a sobering catalogue of first-contact scenarios between technologically disparate societies, in which the less advanced culture was invariably disrupted, displaced, or destroyed. The comparison is not perfect—alien civilisations would not share our biology or historical dynamics—but the asymmetry of power remains a chilling constant. If we cannot guarantee the safety of contact, critics argue, we should not take the risk.

Astrophysicist Stephen Hawking famously warned against active SETI, suggesting that alien civilisations might regard Earth as we do microbial life: unremarkable, disposable, or useful only for exploitation. The danger, he implied, is not necessarily malevolence, but indifference. An advanced civilisation could act in its own interests with consequences we cannot anticipate. The very act of revealing ourselves may collapse the informational advantage we now enjoy—our anonymity.

Moreover, the epistemological assumption that communication is possible across civilisational or biological gulfs may itself be flawed. Critics point out that we have no evidence that alien life would share any of the symbolic systems upon which our messages depend—language, mathematics, or even the logic of reciprocal contact. The risk, then, is not only existential but futile: shouting into a void not just of silence, but of conceptual incoherence.

Another layer of critique revolves around the ethics of consent. METI, by its very nature, is a species-wide act—but it is rarely treated as such. Who has the authority to speak for Earth? Who decides what to say, how to say it, and when to send it? These are not technical questions; they are profoundly political. In the absence of a global framework, decisions are being made by small, often unaccountable groups. The Arecibo message, for example, was crafted and transmitted without public consultation. More recent projects—like the Cosmic Call messages sent in the late 1990s and early 2000s—were similarly unilateral. These initiatives, however well-intentioned, raise uncomfortable parallels with colonial-era declarations made on behalf of populations that had no voice in the matter.

There are, of course, nuanced positions between enthusiastic advocacy and outright prohibition. Some suggest that any METI effort should be preceded by robust international discussion and ratification, ideally under the auspices of the United Nations. Others argue for a tiered approach, beginning with low-information, non-provocative signals designed more to detect than to reveal. Still others suggest a focus on non-directional, time-encoded signals—messages that unfold slowly, giving both sender and recipient time to evaluate and calibrate. These middle paths recognise that while contact may be inevitable in the long arc of cosmic history, its timing and nature are matters of profound consequence.

The debate is also complicated by the fact that Earth has already been "leaking" signals into space for more than a century—radio, television, radar. Some argue that the horse has already bolted, and that active METI is simply a more focused, deliberate extension of what has already occurred. But this argument obscures an important distinction. Incidental radiation is diffuse, rapidly attenuating, and not directed. METI, by contrast, is targeted, high-powered, and designed for detectability. It is not a passive by-product of civilisation—it is a deliberate invitation.

At its heart, the METI debate is not just a question of science or policy. It is a meditation on human identity. Are we a species defined by caution, or by curiosity? Do we act as guardians of our own survival, or as emissaries of our culture? Are we prepared to accept the consequences—positive or negative—of contact? And perhaps most poignantly, are we willing to embrace the possibility that our message may never be received, and that the act of sending is its own form of meaning-making?

There is an aesthetic and philosophical dimension to METI that is often overlooked. To transmit a message into the void is to engage in an act of cosmic theatre—a gesture that may be heard, misunderstood, or ignored, but that nevertheless declares our presence with dignity and hope. Like cave paintings left for descendants, or music composed for an unknown audience, METI is a testament to our desire not only to be known, but to be known as we are: flawed, aspiring, complex, and curious.

Whether we ultimately choose to transmit or to wait, the debate around METI will continue to evolve. It is not a binary contest between recklessness and cowardice. It is a multidimensional dialogue—one that reflects the full range of human thought, from pragmatism to wonder, from vigilance to yearning. In facing this question, we are not just contemplating the universe. We are confronting ourselves. And in that reflection lies the real message we may be sending, whether we press the button or not.

What Message Do We Send—And Who Decides?

The prospect of interstellar messaging raises a question as profound as it is unresolved: if we were to send a message into the cosmos—one that might serve as humanity's first impression to another intelligent civilisation—what should we say? This question, deceptively simple, opens a complex discourse at the intersection of philosophy, politics, ethics, and representation. The issue is not only about the content of the message but also about legitimacy: who is authorised to speak on behalf of humanity, and by what mandate? This debate forces us to confront the fractured nature of global governance, the cultural specificity of values, and the risks of projecting a narrow vision of Earth under the guise of universality.

The history of interstellar messaging offers a useful starting point. In 1974, the Arecibo message—crafted by Frank Drake, Carl Sagan, and colleagues—was transmitted towards the globular star cluster M13. Composed in binary, it encoded information about our DNA, solar system, and basic human form. While technically impressive, the message reflected the biases of its creators: an English-speaking, scientifically oriented, Western academic team. Similar critiques have been directed at the Voyager Golden Record (1977), which included greetings in 55 languages, music from various cultures, and images curated to showcase the diversity of life on Earth. Yet even this record, celebrated as a triumph of collective expression, was the product of a small, unelected group working under significant time and institutional constraints.

What these early efforts reveal is that attempts to "speak for humanity" often fall short of representing its full plurality. Earth is not a unified polity. It is a mosaic of cultures, beliefs, languages, and histories—many of which have been marginalised or excluded from global narratives. To transmit a message as if it encapsulates the human experience inevitably raises issues of epistemic authority and cultural hegemony. Who decides which voices are included? Whose values are foregrounded? What does it mean to speak as "we" in a world still divided by deep inequities?

Consider, for instance, the tension between scientific rationalism and spiritual cosmologies. Should a message prioritise mathematics as a "universal language," or should it include myth, metaphor, and subjective meaning? Should it depict Earth as a technological civilisation, a planetary biosphere, or a sentient species still wrestling with its own ethical contradictions? These choices are not trivial—they encode implicit assumptions about what intelligence is, what matters, and what is worth sharing. A message that privileges technological prowess may inadvertently frame humanity as a species defined by tools and conquest, while neglecting the emotional, artistic, or relational dimensions that many would argue are equally central to our identity.

There is also the challenge of temporality. Any message we send today may not be received for decades, centuries, or even millennia. It may arrive in a future where our own species has evolved, declined, or transformed beyond recognition. To speak for humanity, then, is not only to represent the present but to anticipate how we wish to be remembered—or imagined. This forward-looking orientation requires humility. Are we offering a snapshot, a confession, an invitation, or a warning? Each carries its own rhetorical weight, and each implies a different vision of who we believe ourselves to be.

The politics of representation become even more charged when we consider the actors currently driving METI initiatives. Unlike the multilateral agreements that govern other global commons—such as the Outer Space Treaty for space exploration or the Paris Agreement for climate—there is no international legal framework for messaging extraterrestrial intelligence. This vacuum has allowed individual researchers, private organisations, and even wealthy enthusiasts to pursue interstellar communication without broader consultation. The lack of governance is not merely procedural; it is ethical. In a world where geopolitical tensions persist, where indigenous epistemologies are often sidelined, and where the digital divide remains stark, it is difficult to argue that any single group can credibly claim to speak for "humanity" as a whole.

Efforts to address this gap have been made, albeit sporadically. In 2010, a workshop convened by the Royal Society in London brought together scientists, philosophers, and anthropologists to discuss the implications of alien contact. The participants called for inclusive deliberation before any message is sent, recommending the involvement of global civil society. Similarly, the SETI Institute and other bodies have proposed guidelines that emphasise transparency, multilateral review, and ethical oversight. Yet none of these proposals has led to binding international protocols, and none has successfully mobilised a truly planetary democratic process.

So, what might a more representative approach look like? It would begin by recognising that communication is not a technical artefact but a cultural act. Rather than compressing humanity into a few data points or symbolic gestures, we could embrace a pluralistic model—one that includes competing narratives, diverse cosmologies, and conflicting emotions. The message could be layered, multivalent, or even self-reflexive: a signal not of certainty, but of curiosity, contradiction, and becoming. Such an approach might not be "efficient" in a technical sense, but it would be honest—and perhaps more intelligible to a civilisation equally shaped by its own diversity and dilemmas.

There are practical precedents for such pluralistic endeavours. The World Digital Library, a UNESCO project, curates texts and artefacts from across the globe with the explicit aim of fostering intercultural understanding. The Earth Charter, drafted in consultation with thousands of individuals from all continents, outlines shared ethical principles for sustainability and peace. While these initiatives do not speak with a single voice, they speak with depth, nuance, and legitimacy. A message to extraterrestrials modelled on such frameworks would acknowledge the fragmented, evolving nature of our civilisation—and might be more comprehensible than a contrived narrative of coherence.

Ultimately, the question of what message to send cannot be disentangled from the question of who we are. In choosing what to say, we are forced to confront the ways in which we construct our collective identity. Are we emissaries of knowledge, survivors of conflict, or custodians of a fragile planet? Do we see ourselves as unified by biology, divided by belief, or connected by a shared longing for meaning? These are not questions with simple answers, but the act of grappling with them may itself be the message. Perhaps what we most need to convey is not the illusion of certainty, but the authenticity of struggle.

To speak for humanity is not merely a linguistic challenge—it is a moral one. It demands not only technical sophistication, but political legitimacy, cultural sensitivity, and philosophical courage. It requires us to listen to one another before we speak to the stars. And in that listening, we may discover that the real value of METI is not what we transmit, but how the effort reshapes our understanding of ourselves. For in attempting to speak with the Other, we are inevitably confronted with the question of what it means to speak as "us." That reckoning, however difficult, may be the most important message we ever send.

Assuming Peace: A Dangerous Bias?

In the long arc of human history, contact between distinct civilisations has rarely been neutral, let alone harmonious. Encounters have typically been shaped by asymmetries of power, misunderstandings, and the drive to extract, dominate, or convert. And yet, when it comes to contemplating contact with extraterrestrial intelligence, many within the scientific and cultural mainstream assume that it will be peaceful—imbued with mutual respect, driven by shared curiosity, and unfolding with benevolent intentions. This assumption, while comforting, is anything but neutral. It is a philosophical gamble that overlays our hopes upon a cosmic unknown. And like all unexamined premises, it may blind us to the range of possibilities that contact could entail—including those that are indifferent, predatory, or ethically alien.

The expectation of peace is not accidental. It reflects a deeply rooted Enlightenment ideal: that intelligence and morality are naturally aligned, that reason begets compassion, and that technological advancement implies ethical maturity. Carl Sagan, one of the most prominent advocates for the search for extraterrestrial life, famously suggested that any civilisation advanced enough to engage in interstellar communication must have outgrown war and aggression, or else it would have destroyed itself long before attaining such capacities. This reasoning has an elegant internal logic, but it rests on a set of anthropocentric projections—that intelligence evolves in patterns similar to ours, that survival requires moral development, and that empathy is a universal value rather than a contingent one.

Such optimism is not without merit. It may well be the case that cooperative behaviour, long-term thinking, and conflict resolution are prerequisites for the sustainability of advanced societies. The very act of investing in interstellar communication suggests a capacity for abstract thought, delayed gratification, and a non-zero-sum worldview. But these qualities do not guarantee peace, nor do they exclude alternative modes of survival and expansion. On Earth, many of the most adaptive species—both human and non-human—have thrived not through cooperation alone, but through competition, deception, and exploitation. Evolution favours fitness, not fairness. And if this principle holds elsewhere, then intelligence may coexist with behaviours we would consider amoral, or even malevolent.

One of the most forceful critiques of the peace assumption has been articulated through the "Dark Forest" hypothesis, popularised by Chinese science fiction author Liu Cixin. In this view, the universe is not a cosmic marketplace of benevolent actors, but a cold, silent expanse where every civilisation hides from the others out of fear. The logic is chillingly simple: you cannot be certain of another civilisation's intentions, and by the time you know, it may be too late to defend yourself. Therefore, the safest strategy is to remain silent—or to eliminate potential threats pre-emptively. While speculative, this framework is grounded in game theory and realpolitik, and it echoes strategies familiar to those who study international relations or military deterrence.

Indeed, when viewed through the lens of history, the assumption of peaceful contact appears dangerously naive. When technologically superior societies on Earth encountered others, the results were often catastrophic. The arrival of Europeans in the Americas led not to cultural exchange, but to genocide, disease, and dispossession. Colonial encounters in Africa, Asia, and the Pacific were justified through narratives of civilisational superiority, often couched in the language of benevolence. These precedents caution against assuming that difference will be met with respect, or that superiority will be tempered by empathy.

A more rigorous approach requires acknowledging that alien intelligence, if it exists, may not share our moral intuitions, emotional architecture, or ethical norms. Even the concept of "benevolence" may be meaningless or distorted when applied to non-human cognition. What if an alien civilisation understands life as a resource to be optimised, rather than a sanctity to be preserved? What if its form of empathy extends only to entities that share its substrate—be it silicon, plasma, or quantum field? What if it perceives communication not as dialogue, but as dominion? In such cases, our attempts at contact may be interpreted not as friendly gestures, but as vulnerabilities to be exploited—or irrelevancies to be dismissed.

This is not to say that we should assume hostility, but rather that we should approach contact with epistemic humility and strategic caution. The anthropologist's principle of cultural relativism—the idea that one must understand practices in terms of their own context—may serve as a useful analogue. We should not expect aliens to conform to our norms, nor should we presume to be legible to them. Even the concept of "intelligence" may be broader and more fractured than we imagine. A sufficiently advanced civilisation may operate through distributed minds, hive consciousness, or modes of perception that render our signals incomprehensible or annoying. The danger lies in assuming that our gestures of goodwill will be received as intended—and in failing to consider the possibility that misunderstanding could have irreversible consequences.

Real-world initiatives already reflect this dilemma. The METI project, which advocates for actively sending messages into space, has drawn sharp criticism from scientists like Stephen Hawking and David Brin, who warn that revealing our presence may be unwise. Unlike SETI, which listens passively, METI involves deliberate transmission—an act that assumes not only the existence of receivers, but that those receivers will interpret our signals as non-threatening. Critics argue that even if the risk is low, the potential cost—existential harm—is too great to justify proceeding without global consensus. In other words, the bias towards peace is not merely philosophical; it has policy implications with planetary stakes.

Some have suggested that a more prudent approach would be to develop layered messaging protocols, starting with low-information signals—mathematical patterns, astronomical coordinates, or representations of natural laws. These would serve as probes, testing whether there is a response and how it manifests. Only after such an exchange proves stable should more detailed information be transmitted—about our biology, culture, or technologies. This incremental strategy mirrors diplomatic practices, in which trust is built through small, reversible gestures before deeper engagement occurs. It also reflects a fundamental insight: that peace is not the default state of contact, but an achievement that must be cultivated through care, reciprocity, and restraint.

There is also a psychological dimension to the peace assumption that merits scrutiny. It may reflect not only hope, but denial. In imagining aliens as wise and benevolent, we externalise our longing for guidance, redemption, or validation. The alien becomes a cosmic parent, teacher, or saviour—an omniscient figure who can solve our problems or affirm our worth. This narrative, while compelling, obscures the fact that we are alone in our decisions—for now—and that our future depends on how we confront uncertainty, not on how we imagine rescue.

Ultimately, the question is not whether alien civilisations are likely to be peaceful, but whether we are prepared for the full spectrum of possibilities. To assume peace is to project our best qualities onto an unknown other. To prepare for risk is to acknowledge the limits of our understanding and the fragility of our position. These positions need not be mutually exclusive. We can retain the hope for contact while recognising that our approach must be shaped by caution, not complacency.

Assuming peace is not a harmless ideal—it is a hypothesis with real-world implications. It affects how we design messages, how we engage in policy, and how we think about the future of our species. A more responsible stance requires us to test that hypothesis critically, to build diverse and inclusive frameworks for decision-making, and to remember that in speaking into the cosmos, we are also speaking to ourselves. What we assume about others reveals what we believe about our own nature. And perhaps, in questioning those assumptions, we may discover a wiser way to listen before we speak.

The Morality of Discovery: Ownership, Rights, and Responsibility

The prospect of discovering alien life has long stirred the imagination of scientists, philosophers, and storytellers alike. Yet within this exhilarating prospect lies a quieter, more profound question—one that is ethical rather than empirical: if we were to find life beyond Earth, what rights would it have? Would we see it as an object to study, a resource to exploit, or a sentient peer to be respected? The morality of discovery is not just about what we can do with alien life, but what we *should* do. It calls for a reevaluation of how we understand personhood, dignity, and the responsibilities that come with knowledge and power.

The issue begins with a premise often taken for granted—that discovery confers authority. Historically, the act of "discovery" has been inseparable from claims of ownership. When European explorers encountered new lands and peoples during the Age of Exploration, they often interpreted their arrival as a mandate to possess, categorise, and dominate. This logic, underpinned by the so-called "doctrine of discovery," was used to justify colonial expansion, resource extraction, and cultural erasure. What was "discovered" was treated as if it had no inherent claim to autonomy until recognised by the discoverer. This troubling legacy warns us that the language of discovery is never neutral; it is bound up with assumptions of entitlement, hierarchy, and legitimacy.

Should alien life be found—whether as microbial traces beneath the ice of Europa or in the form of a complex ecosystem on a distant exoplanet—the same philosophical tension will re-emerge. Who gets to define its significance? Who sets the terms of engagement? And crucially, what ethical principles will guide our response? These are not hypothetical questions. They are preemptive ethical imperatives, demanding forethought and humility before we face the reality they portend.

At one end of the spectrum is the reductionist view: that alien life, unless demonstrably intelligent, holds no moral status beyond its instrumental value. From this standpoint, life is data. A Martian microbe is not a peer but a sample. Its DNA—or its alien equivalent—is a code to be sequenced, studied, and possibly commercialised. This utilitarian logic mirrors how humans have often approached life on Earth: valuing organisms for their pharmacological potential, ecological services, or symbolic capital, rather than their intrinsic worth. In such a framework, alien life becomes property by default—catalogued, patented, perhaps even weaponised.

Yet this approach is increasingly out of step with emerging ethical paradigms. Across disciplines, there is a growing recognition of the rights of non-human entities. Legal frameworks have been proposed—and in some cases enacted—to grant personhood to rivers, forests, and ecosystems. The Ecuadorian constitution, for instance, recognises the rights of nature. In New Zealand, the Whanganui River is legally recognised as a living entity with rights. These shifts suggest a broader movement toward recognising the moral status of beings not on the basis of intelligence or utility, but relationality and integrity. If we are beginning to grant rights to non-human life on Earth, should alien life be any less protected?

The question becomes more acute if the life we discover is intelligent or sentient. What if we encounter creatures with language, social structures, or self-awareness—attributes we typically associate with moral status? Would we be prepared to recognise their sovereignty, to treat them not as subjects of study but as holders of rights? And if so, what rights? The right to autonomy? To be free from interference? To be heard, even if we cannot fully understand them?

Our response will hinge, in part, on how we define life's moral significance. Western philosophy has often anchored moral status in rationality—a legacy of Enlightenment thinking that privileges cognition over embodiment, and humans over all else. But this anthropocentric model is increasingly being challenged. Feminist, indigenous, and posthumanist perspectives argue for a more expansive understanding of moral community—one that includes the non-human, the more-than-human, and the radically other. These frameworks are not merely inclusive; they are responsive to the interconnectedness of life. They suggest that discovery entails not dominion, but relationship—and that relationship brings with it ethical obligations.

These obligations are not abstract. They will shape policy, research protocols, and international law. If we detect biosignatures in an exoplanetary atmosphere, should we attempt contact immediately, or proceed with caution? If we land on a moon with microbial life, should we collect samples or quarantine the environment? And who decides? At present, space governance is patchy at best. The Outer Space Treaty of 1967 prohibits claims of national sovereignty over celestial bodies, but it says little about the moral status of life. Existing planetary protection guidelines focus on preventing contamination—not out of respect for alien life, but to preserve scientific integrity and avoid false positives.

To address these gaps, some ethicists have proposed a planetary protection framework rooted not in anthropocentric utility but in bioethical principles. This would treat alien ecosystems—regardless of their complexity—as worthy of moral consideration. It would extend the precautionary principle beyond Earth, arguing that the absence of full knowledge should not be a licence for exploitation. This is not sentimentality—it is prudence. Once a unique form of life is disturbed, it cannot be un-disturbed. The act of observation is already an act of interference.

The implications extend to questions of cultural responsibility as well. Would we export our conflicts, ideologies, or market logics to new worlds? Would the first human representatives of Earth act as stewards or extractors? These are questions not just of ethics, but of planetary maturity. How we treat alien life may be the truest measure of our own civilisation's values. If we respond with care, restraint, and reciprocity, we affirm that our intelligence is accompanied by wisdom. If we respond with greed or indifference, we risk repeating the worst patterns of our past—this time on a cosmic scale.

There is also a psychological component to the morality of discovery. To encounter alien life is to be confronted with the unfamiliar—not just biologically, but morally. It challenges our anthropocentric narratives and forces us to revise our place in the cosmos. This discomfort can be productive. It can invite humility, curiosity, and a willingness to be transformed by encounter. But only if we are prepared to see the "other" not as a reflection of our own fears or fantasies, but as a presence in its own right, deserving of respect.

Respect, in this context, means more than passive non-interference. It implies an active commitment to protect the integrity of the life we encounter. It means accepting that discovery is not a claim to ownership, but a call to responsibility. And responsibility is not something we defer to policy-makers or technocrats—it begins with how we, as a global community, choose to frame the meaning of contact.

As we stand on the threshold of potential discovery, we must ask not only what lies beyond Earth, but what lies within us. Will we carry forward the logic of conquest, or can we imagine a new paradigm—one rooted in mutual respect, interdependence, and moral imagination? The answers to these questions will not be found in telescopes or algorithms, but in ethics, philosophy, and collective deliberation.

The morality of discovery demands that we think beyond the thrill of finding alien life. It asks us to consider what kind of civilisation we aspire to be when we encounter the unknown. Whether the life we find is microbial or sentient, fragile or formidable, its value cannot be measured solely by our curiosity. Its rights—and our responsibilities—must be defined not by the act of finding, but by the wisdom to recognise that some things, once found, must be treated not as trophies, but as neighbours. And in that recognition may lie the first step towards a truly interstellar ethics.

Contact Simulations: What Would You Say?

Imagine this: a narrow-band radio signal arrives from a distant star system, structured with unmistakable markers of non-randomness—mathematical sequences, prime number patterns, or modulated bursts resembling language. After years of speculation, decades of passive listening, and centuries of cultural anticipation, humanity is no longer alone. The first message from an extraterrestrial intelligence has arrived. The world holds its breath. And then, the inevitable question: how do we respond?

This question, while seemingly hypothetical, serves as a crucible for the deepest aspects of human identity, ethics, and imagination. It is not only a test of scientific and linguistic ingenuity but a philosophical litmus test of who we believe ourselves to be. The message we send in return—if we choose to send one—would encapsulate not merely facts about our species or planet, but a self-portrait etched into the stars. What would we say? What *should* we say? And who gets to decide?

Contact simulations offer a powerful tool for engaging with these questions. By constructing interactive scenarios—thought experiments where different messages are proposed and different consequences unfold—we can better anticipate the complexities of interstellar communication and reflect on the values and assumptions that shape our decisions. These simulations are not predictions; they are provocations, designed to illuminate possibilities and challenge complacency.

Consider one such scenario. In Simulation A, humanity chooses to respond with a message focused purely on scientific knowledge. The transmission includes information about our physical laws, chemical elements, mathematical constants, and the structure of DNA. The message is composed in what we hope is a universal language—mathematics—intended to signal our intelligence without revealing our emotional or political characteristics. In this simulation, the alien civilisation responds positively, acknowledging our data with an even more complex reply. But they remain silent on cultural or ethical topics, creating a communication that resembles a technical exchange between computers rather than a meaningful dialogue. The result: contact is established, but the relationship remains sterile, devoid of mutual understanding or emotional resonance. Have we made contact—or merely proven our capacity to calculate?

Now shift to Simulation B. This time, the human message includes cultural artefacts—music, visual art, short narratives, and poetry translated into symbolic code. We transmit Beethoven, Bharatanatyam, Bashō, and Beyoncé. We share images of children, animals, and landscapes from across the globe. The message is a collage of human life, intended not only to inform but to connect. The aliens respond with what appears to be an emotional analogue—modulated frequencies that show variation in rhythm, tone, and perhaps affect. Interpreting the response is difficult; we are unsure whether they are replying in kind or merely echoing our formats. Still, the exchange opens new dimensions: not just informational, but relational. But this scenario introduces new risks—what if some parts of our culture offend or alarm? What if the way we portray ourselves invites misinterpretation? In trying to be understood, do we risk being misunderstood?

Simulation C introduces a strategic framing. The message is sent only after extensive international deliberation, coordinated by an agency governed under a new global protocol for interstellar communication. The content is carefully negotiated: it includes basic scientific knowledge, a statement of peaceful intent, and an invitation for reciprocal dialogue. Religious, political, and commercial messaging are excluded. The simulation models a long delay in response—possibly centuries—but the message is archived and preserved as a formal record of humanity's intentions. This simulation foregrounds diplomacy over spontaneity. It is cautious, measured, and inclusive. But it also faces critique: does it truly reflect the vibrancy and diversity of human life, or does it flatten us into consensus? In trying to avoid controversy, do we sacrifice authenticity?

The final thought experiment—Simulation D—is less optimistic. In this version, a private group of enthusiasts sends a message independently, using powerful directional signals without international oversight. The content is eclectic and emotionally raw—part manifesto, part confession. It includes philosophical musings, warnings about our species' mistakes, and an open plea for guidance. The message goes viral online but draws sharp condemnation from scientific and governmental bodies. Years later, a reply is received—but it is cryptic, ambiguous, and possibly threatening. Were we naïve? Did we invite a form of contact we are not ready to interpret or withstand?

Each of these simulations points to a central insight: what we say to alien intelligence will reflect not only our aspirations but our self-understanding. The message we send is not simply *about* humanity—it *is* humanity, distilled into a signal. And yet, no message can be fully comprehensive, or universally representative. Any attempt to speak for Earth is necessarily partial. It is this partiality that gives contact simulations their ethical power—they reveal that the act of messaging is not just technical but moral.

Historical case studies on Earth further enrich this reflection. When the Polynesians first made contact with European sailors, the initial exchange of artefacts and gestures quickly gave way to conflict, misunderstanding, and exploitation. The arrival of European ships on the shores of Africa, Asia, and the Americas also began with messages—flags, offerings, gifts—but ended in colonisation and collapse. These terrestrial precedents caution us that the message matters less than the relational context in which it is interpreted. Communication is always embedded in power, perception, and asymmetry.

The content of a message must therefore be accompanied by an understanding of its possible reception. We must ask not only what we are saying, but how it might be read—or misread. A depiction of Earth's weaponry might be interpreted as aggression or insecurity. A portrayal of global poverty might evoke pity, contempt, or opportunism. A celebration of diversity might be seen as incoherence. Conversely, a carefully curated image of utopian harmony might come across as deceptive or insincere. Every inclusion and omission sends a signal.

This is why interactive simulations—designed to model responses to different message types—are more than academic exercises. They help us think through the communicative act as a system of meaning, not just data. They invite interdisciplinary engagement, bringing together linguists, diplomats, anthropologists, ethicists, and artists alongside scientists and engineers. They shift the question from "what can we say?" to "what should we say—and why?"

Such simulations are already being explored in academic and policy contexts. The "Contact Project" initiated by SETI researchers and social scientists aims to develop protocols for post-detection communication. Scholars like Douglas Vakoch and Kathryn Denning have argued for more inclusive, culturally aware approaches to messaging, emphasising pluralism over precision. Emerging projects even allow public participation, enabling citizens to submit potential messages and vote on their contents—a democratisation of messaging that reflects the distributed nature of our species.

But these efforts must also confront their limits. No simulation can fully model the psychology or semiotics of alien life. We are guessing across ontological chasms. The exercise is valuable not because it prepares us for certainty, but because it habituates us to uncertainty—to the ethics of making decisions under profound ambiguity. The process of simulating contact is, in effect, a rehearsal for humility.

Ultimately, the power of these thought experiments lies in their capacity to destabilise comfortable narratives. They remind us that contact is not just a scientific milestone—it is a civilisational mirror. What we choose to say will reveal not only our intelligence, but our priorities, our fears, and our capacity for empathy. It will reveal whether we see ourselves as isolated individuals, rival nations, or a single planetary species poised—tentatively—to reach across the void.

So what would *you* say?

Would you offer a poem or a prime number? A peace treaty or a confession? A whisper or a song?

In asking this, we do more than imagine the stars—we begin to imagine ourselves anew. And perhaps, even before the first alien reply arrives, that act of imaginative reflection is its own kind of contact.

Chapter 8
Religion, Myth, and Alien Revelation

Since the dawn of civilisation, humanity has looked skyward for answers, comfort, and meaning. The stars have not only served as navigational aids or calendars, but as canvases upon which cultures projected their deepest fears and highest aspirations. Gods, ancestors, spirits, and messengers have populated the heavens in myths that span continents and millennia. But what if some of those ancient stories were rooted not only in metaphor, but in encounters—misunderstood or mythologised—with intelligences from beyond Earth? And more provocatively, what happens to our religious and mythological frameworks if we confirm the existence of life elsewhere in the universe?

This chapter explores a rich and delicate intersection: the encounter between the cosmic unknown and the sacred known. It considers how the discovery of extraterrestrial life—whether microbial or intelligent—might destabilise, reshape, or reaffirm our religious and mythic narratives. Far from being marginal to the scientific discourse on alien life, religious traditions and mythologies are central to how billions of people understand creation, purpose, and human exceptionalism. Thus, any serious contemplation of contact must reckon not only with biological implications but theological ones.

We begin with a long-standing cultural fascination: the blurred boundary between ancient myth and modern speculation. The so-called "ancient astronaut theory," popularised in the 20th century by authors like Erich von Däniken, claims that many of humanity's foundational myths—gods descending from the sky, divine knowledge imparted to mortals, miraculous technological feats—are evidence of prehistoric alien contact. These narratives are often dismissed by historians and archaeologists for their lack of empirical support and their Eurocentric, sometimes colonialist, assumptions about the capabilities of ancient peoples. Yet their popularity points to a deeper yearning: to reconcile myth with science, the sacred with the speculative. Whether these stories are factual or not is less important than what they reveal about our cognitive instinct to populate the unknown with meaning-bearing beings.

Throughout recorded history, different cultures have told remarkably similar tales of beings from the heavens: Sumerian Anunnaki, Hopi kachinas, Hindu devas and asuras, the Nephilim of Genesis, or the Dogon sky visitors in Mali. While scholars rightly caution against reading these traditions through the anachronistic lens of modern ufology, they also note that these stories often emerged in response to phenomena that felt beyond explanation. In this sense, the alien has always been with us—not as an empirical presence, but as a narrative archetype. The possibility of actual extraterrestrial life, then, threatens not so much to displace myth as to crystallise its subtext: the longing for cosmic context and relationality.

But myth and religion are not synonymous, and it is organised religion that may face the most immediate existential shock if alien life is confirmed. This is particularly true of the monotheistic Abrahamic faiths, which traditionally present humanity as central to creation. In Christianity, for instance, the Incarnation—the idea that God became human in the person of Jesus—is profoundly anthropocentric. Would this event have occurred on multiple planets, in different forms? Would Christ be incarnated as a different species in another world? Or is humanity uniquely chosen, while other intelligent beings remain outside salvation history? Such questions, once the domain of theological fiction, may become urgent doctrinal challenges.

The Vatican has already begun preparing for this possibility. Prominent Jesuit astronomers affiliated with the Vatican Observatory, such as Brother Guy Consolmagno, have publicly stated that the discovery of alien life would not necessarily contradict Catholic theology. Consolmagno has suggested that extraterrestrials, like humans, could be "creatures of God" and part of a broader divine plan. Similar openness can be found among some Islamic scholars, who note that the Qur'an speaks of "beings of the heavens and the earth" and explicitly allows for the existence of other sentient creations beyond humans and jinn. In contrast, more fundamentalist interpretations—across religious traditions—may struggle with the ontological implications of non-human intelligence, especially if it exhibits moral reasoning, creativity, or spiritual practice.

Eastern philosophies may offer a different paradigm. In many Hindu, Buddhist, and Taoist cosmologies, life is not confined to Earth. Multiple realms, cycles of rebirth, and interconnected planes of existence already accommodate the idea of otherworldly beings. In this view, the discovery of alien life might not be disruptive but integrative—fitting naturally into an already pluralistic metaphysical vision. Such systems remind us that not all religious traditions are equally vulnerable to theological destabilisation; some may even find their cosmologies affirmed.

Still, the question remains: how do we craft a spiritual framework that accommodates both scientific discovery and existential significance? The term *cosmotheology* has emerged as a way to describe new theological efforts that incorporate the possibility of a populated universe. Thinkers like Ted Peters, Paul Tillich, and Teilhard de Chardin have envisioned a theology that expands the sacred horizon to include alien life. In such models, God is not the deity of one planet but the ground of being across the cosmos. The Incarnation, under this view, may be one expression of divine engagement among many. Alien life, rather than threatening faith, becomes another expression of divine creativity.

This reframing opens the door to what we might call *astroethics*: the responsibility to engage alien life not only scientifically but morally. If we encounter intelligent beings, do they have souls? Do they seek transcendence, ask about purpose, or venerate the numinous in their own ways? Would we recognise their rituals, myths, or sacraments? Or would we project our frameworks onto them—misreading silence as emptiness, or difference as deficiency? The risk of cultural imperialism extends beyond Earth. If we are to be spiritual ambassadors, we must proceed with humility, not missionary zeal.

But not all reactions to alien contact would be utopian. Some theologians, psychologists, and cultural theorists warn that the discovery of advanced alien life could trigger a spiritual crisis. For those whose religious identity is tied to human uniqueness, contact might induce existential despair. It could be read as evidence that humanity is not special, or that the gods we revere are local, not universal. This anxiety may manifest not only in religious sects but in secular existentialism: a renewed confrontation with the cosmic indifferentism that thinkers like Camus and Sartre described. The discovery of alien life could be our generation's Copernican trauma—a reshuffling of our place in the hierarchy of meaning.

And yet, as history shows, existential disorientation can also be generative. The Copernican revolution did not destroy faith; it provoked theological adaptation. Darwinian evolution did not end religion; it compelled deeper inquiries into purpose and design. In the same way, the confirmation of extraterrestrial intelligence might catalyse a spiritual renaissance, forcing us to rewrite—not abandon—our origin stories.

Such stories are not fixed. They evolve, absorb, and reconfigure. Already, writers, artists, and theologians are imagining origin narratives that include alien life: the seeding of Earth by panspermia; the guiding hand of alien civilisations in our development; or parallel spiritual evolutions across star systems. These speculative frameworks are not delusions—they are acts of cosmological imagination. They remind us that meaning is not destroyed by scale; it is expanded by context.

Ultimately, this chapter invites the reader to consider that the discovery of alien life may be less about displacement and more about integration. It may not answer our deepest questions—but it will change the questions we ask. It may not confirm our beliefs—but it will refine them. In this sense, the true revelation may not be about the aliens at all. It may be about us: how we confront mystery, how we adapt our myths, and how we re-enchant a universe that is suddenly, unimaginably alive.

Ancient Aliens vs. Ancient Myths: Sorting Fact from Folklore

The notion that ancient myths conceal traces of extraterrestrial encounters has captured the popular imagination for decades. Television series, bestselling books, and online forums brim with claims that gods, angels, and sky beings described in the sacred texts and oral traditions of antiquity were, in fact, misunderstood visitors from other worlds. Yet beneath the spectacle of pseudo-history lies a more nuanced and intellectually vital inquiry—how do cultures use myth to explain the unexplainable? And could ancient stories, stripped of literalism, still bear witness to profound cosmic encounters, whether real or metaphorical?

At the heart of this debate lies a fundamental tension between symbolic meaning and empirical evidence. On one side stands the "ancient astronaut theory," which posits that many ancient structures, artworks, and myths are evidence of advanced alien intervention. On the other stands the scholarly consensus: that these myths are expressions of spiritual insight, social order, and cosmological awe, not covert records of interstellar visitation. Yet to dismiss the ancient alien hypothesis outright, without unpacking its psychological and cultural appeal, would be to miss the deeper point—these ideas persist not because they are necessarily true, but because they speak to something unresolved in the human psyche.

Let us begin with the core claim: that extraterrestrials visited Earth in antiquity and were worshipped as gods. Proponents of this idea frequently cite the monumental architecture of ancient Egypt, Mesopotamia, and Mesoamerica as evidence of technological input from beyond the stars. The Great Pyramid of Giza, the Nazca Lines of Peru, and the colossal stone heads of Easter Island are all said to be beyond the capabilities of the societies that built them. Similarly, mythological accounts of flying chariots, glowing beings, and celestial battles are reinterpreted as eyewitness testimonies to alien contact, encoded in the language of myth because the witnesses lacked the vocabulary to describe what they saw.

A closer examination, however, reveals the intellectual flaws in such arguments. Archaeologists, engineers, and anthropologists have extensively documented how these structures could have been built using the tools and knowledge available at the time. The logistics of transporting stone, the use of astronomical alignments, and the symbolic functions of monumental design are well understood. Suggesting otherwise often reflects an implicit—and troubling—bias: that non-Western societies could not have achieved these feats without outside help. This not only undermines the ingenuity of ancient civilisations but perpetuates a neo-colonial view of history that devalues indigenous knowledge.

Yet the persistence of the ancient alien hypothesis is not merely a failure of education—it is a reflection of existential curiosity. Myths, by their nature, are vehicles for meaning. They are not journalistic accounts but symbolic narratives that encode moral lessons, metaphysical insights, and cosmic frameworks. When ancient texts describe beings descending from the sky, or fiery wheels within wheels, or gods who traverse the heavens in vessels of light, they are not necessarily reporting events—they are expressing awe, fear, wonder, and a desire to situate the human experience within a larger cosmic order.

Consider, for example, the Sumerian myth of the Anunnaki, a pantheon of deities associated with the heavens, who allegedly came to Earth and created humankind. In modern alternative theory circles, this myth is often reinterpreted as a literal account of genetic engineering by alien beings. But within its original cultural context, the Anunnaki reflect theological concerns about divine kingship, agricultural cycles, and the social contract between rulers and the gods. Similarly, Ezekiel's vision in the Hebrew Bible—often cited for its "wheels within wheels" imagery—was a religious epiphany, not a field report on UFO sightings. These stories gain power not from their literalism, but from their capacity to articulate the sacred through poetic imagination.

Where the ancient alien hypothesis falters, mythology flourishes. Myths are not failed attempts at science; they are successful attempts at meaning-making. They operate on a different epistemological plane—one where metaphor, analogy, and narrative structure convey truths that transcend empirical verification. Rather than seeing them as garbled history, we might do better to see them as reflections of universal archetypes: the sky as source, the Other as revealer, the gods as mediators between the known and the unknown.

However, this does not mean we should dismiss all possibilities of ancient cosmic encounters. On the contrary, a more intellectually honest approach is to acknowledge the limits of both literalism and scepticism. There remain anomalies in the historical and archaeological record—puzzling artefacts, unexplained alignments, and motifs that recur across cultures separated by geography and time. While these do not prove alien visitation, they do suggest that ancient peoples were deeply attuned to the cosmos, both as a physical environment and as a symbolic realm. Their myths reflect an attempt to engage with mysteries that are, in many ways, still with us.

The real question is not whether aliens built the pyramids, but why we are still drawn to this idea. It is no coincidence that the surge in ancient alien speculation coincided with the space age. As we ourselves became capable of space travel, we began to reimagine the past in light of this new cosmic horizon. The same technological optimism that landed humans on the Moon in 1969 also fuelled the idea that others might have come before us—and perhaps, left clues behind. This backward projection reflects a modern myth: that contact with the Other is both imminent and redemptive.

Moreover, the resurgence of ancient alien narratives speaks to a deep cultural anxiety: that our traditional religious frameworks are no longer sufficient to contain the scale of the universe as revealed by science. In a cosmos of trillions of galaxies, each with billions of stars, the idea that we are alone—and central—feels increasingly untenable. Ancient alien theories, for all their shortcomings, offer a bridge between myth and modernity. They promise that the gods were real, that we are not alone, and that our ancestors were part of a larger cosmic story.

But there is a cost to conflating myth with history. When we reinterpret spiritual symbols as spacecraft, we strip them of their metaphysical power. We reduce the sacred to the technological, the transcendent to the mechanical. In doing so, we risk losing the wisdom encoded in these stories—not as blueprints for machines, but as blueprints for meaning. Myth teaches us how to orient ourselves in a cosmos that is vast, mysterious, and potentially inhabited. It teaches us humility, wonder, and reverence.

Sorting fact from folklore, then, requires more than debunking. It requires listening—carefully and contextually—to what the ancients were really trying to say. Not "aliens came," but "the universe is alive." Not "gods flew spaceships," but "the human soul reaches for the stars." By recovering the symbolic language of myth without reducing it to science fiction, we gain a deeper understanding of ourselves and our perennial search for cosmic connection.

In the end, the fascination with ancient aliens reveals less about the past than it does about the present: our hunger for meaning in a disenchanted age, our desire to belong to something larger, our fear that we might be insignificant—and our hope that we are not. If we are to engage seriously with the question of life beyond Earth, we must be able to distinguish between fantasy and insight, between literalist distortions and symbolic truths. Only then can we honour both the intelligence of the ancients and the curiosity of the modern mind—and prepare ourselves for the real mysteries that may lie ahead.

How the Discovery of Life Would Challenge World Religions

For millennia, the fundamental narratives of the world's great religious traditions have placed human beings at the centre of creation. Whether conceived as divine stewards, spiritual wayfarers, or fallen beings in need of redemption, humanity has been the focal point of sacred cosmologies. But what happens when this anthropocentrism is confronted by the confirmation of alien life—life that is not only biologically other but, potentially, sentient, intelligent, and spiritually aware? The question is not merely speculative. It is an impending theological frontier, and one that demands deep engagement from scholars, clerics, and lay believers alike.

The initial challenge is conceptual: most religious traditions have evolved in cultural and metaphysical contexts where Earth was assumed to be the unique cradle of life. The heavens were spiritual rather than biological spaces—populated by angels, demons, gods, and celestial bodies imbued with symbolic meaning. A living universe teeming with autonomous, evolved beings was never factored into the doctrinal frameworks of these traditions. As such, the discovery of life elsewhere—particularly intelligent life—could provoke theological disruption, cognitive dissonance, or radical reinterpretation, depending on the flexibility of the tradition in question.

Christianity, with its strong emphasis on incarnation, salvation, and the unique status of humanity, faces some of the most complex theological implications. Central to Christian doctrine is the belief that Jesus Christ, the Son of God, became incarnate in human form to redeem fallen humanity. The question then arises: if intelligent beings exist on other worlds, do they share in this fallen nature? Would they require salvation? If so, did Christ's incarnation apply universally, or would God need to incarnate multiple times in different forms across the cosmos?

Some Christian theologians have begun to grapple with these questions in earnest. The Vatican Observatory, under the stewardship of figures such as Brother Guy Consolmagno and Father José Gabriel Funes, has affirmed that the existence of extraterrestrial life would not necessarily contradict Catholic doctrine. In this view, God's creative power is not confined to Earth, and other beings may enjoy their own relationship with the divine. Yet the implications for soteriology—the theology of salvation—remain unsettled. As the Anglican theologian C. S. Lewis once speculated in his space-themed fiction, alien beings might either be unfallen (and therefore not in need of redemption) or have their own salvific narratives unknown to humanity.

Islam may offer a more adaptable metaphysical framework. The Qur'an speaks of "the Lord of the Worlds," a phrase that, taken literally, implies a universe populated with many domains of existence. The existence of other beings is not only compatible with Islamic cosmology but anticipated by certain verses. For instance, Surah 42:29 states, "And among His signs is the creation of the heavens and the earth and whatever living creatures He has dispersed throughout them. And He is competent to gather them when He wills." This passage suggests that life beyond Earth is within the divine plan and does not contradict Islamic teaching. While Islam, like Christianity, holds humanity as a privileged creation, it does not dogmatically assert exclusivity. The door is open, though not yet doctrinally detailed.

Judaism, with its less centralised doctrinal structure, may respond with similar flexibility. Its theological focus is less on universal claims about salvation and more on covenantal relationship—the unique bond between God and the Jewish people. This specificity allows room for the possibility that other beings may have their own covenants or spiritual realities. Rabbinic literature often entertains speculative questions, and some Jewish scholars have argued that if alien life exists, it would simply reflect the vastness of God's creation. The late Rabbi Aryeh Kaplan, for example, considered the existence of extraterrestrial life not only plausible but consistent with Jewish mystical thought, particularly Kabbalistic notions of multiple worlds and cosmic realms.

Hinduism, with its richly pluralistic cosmology, may be the most naturally accommodating to the notion of extraterrestrial life. The ancient Vedas describe numerous realms, inhabited by diverse beings with varying degrees of consciousness and spiritual attainment. Concepts such as the *lokas* (worlds), *devas* (celestial beings), and *asuras* (anti-divine entities) already imply a cosmos that is populated, layered, and hierarchically structured. Life elsewhere in the universe is not an intrusion into the sacred narrative—it is intrinsic to it. Moreover, the idea of reincarnation across different planes of existence allows for a metaphysical continuity that could easily integrate alien life into the spiritual cycle of *samsara* and *moksha*.

Buddhism offers a similarly non-anthropocentric vision. Life is understood as manifesting across multiple realms, including those of animals, humans, hungry ghosts, and celestial beings. The criteria for moral and spiritual development are based on consciousness and karma, not species or biology. From this perspective, extraterrestrial beings with sentience would simply be fellow participants in the cycle of birth, death, and rebirth. The path to enlightenment would remain open to them, just as it is to humans. The discovery of such beings might even be seen as an opportunity to expand compassion and deepen the understanding of *interbeing*—the interdependent nature of all existence.

Yet, despite these philosophical frameworks, practical challenges remain. Institutional religions often move slowly, and the emotional responses of believers may vary widely from the official theological positions of clerics and scholars. A key question is how religious communities would react to a discovery that challenges their sacred narratives. Would they reinterpret their texts symbolically, as many did after the Copernican revolution and Darwinian evolution? Or would they double down on literalism, potentially fracturing into sects or prompting defensive retrenchment?

History offers precedent for both responses. The heliocentric model initially provoked ecclesiastical resistance, yet over time was absorbed into mainstream religious thought. Similarly, evolution was initially decried by many religious leaders but is now accepted by numerous denominations as compatible with the idea of divine creation. In both cases, religious traditions adapted—not by abandoning core beliefs, but by re-contextualising them in light of new knowledge. The challenge posed by alien life may follow this pattern, demanding intellectual rigour, spiritual maturity, and institutional courage.

Moreover, religious thought need not be reactive. It can also be prophetic—capable of anticipating and integrating new realities before they become mainstream. Interfaith dialogue, for instance, could take up the question of cosmic ethics: what moral principles would guide our relationship with alien beings? Would we be bound to extend compassion, justice, or stewardship? Could theological ethics help frame planetary protection, non-interference, or the rights of non-human intelligences? These are not merely academic exercises—they are anticipatory acts of moral imagination.

In some sense, the discovery of alien life offers religion a renewed vocation. Rather than defending outdated anthropocentric models, religious traditions could become facilitators of cosmic humility. They could teach that human uniqueness is not contradicted by other life forms but contextualised by them. They could explore what it means to be made in the image of God when that image may have more than one expression. They could offer rituals, myths, and symbols that help humanity make sense of its place in a populated universe.

Ultimately, how world religions respond to the discovery of alien life will depend on their willingness to distinguish essence from form—core truths from culturally contingent narratives. If the essence of religion is a search for meaning, a framework for morality, and a language for the sacred, then the existence of other life need not threaten it. Instead, it may enrich and extend it. Just as telescopes expanded our view of the cosmos, so too might contact expand our spiritual horizons. And in doing so, religion may fulfil one of its oldest promises: to help us not merely look up at the stars, but understand our place among them.

A New Cosmotheology: The Divine in a Populated Universe

The prospect of a universe teeming with life is not merely a scientific revelation—it is a theological revolution in waiting. For centuries, humanity's religious and philosophical frameworks have grappled with the divine in relation to a singular world: Earth. Even as our telescopes extended our vision across the heavens, the metaphysical maps we carried remained remarkably geocentric. But as the possibility of extraterrestrial intelligence becomes more than speculative fiction—supported by rigorous planetary science and increasingly sophisticated detection technologies—the need arises for a new theological architecture: one spacious enough to house the divine across a cosmic scale, yet coherent enough to preserve spiritual depth. This is the frontier of cosmotheology.

Cosmotheology, in its emerging modern form, is the effort to articulate a spiritual and metaphysical understanding of the divine that remains viable in a universe that may include many worlds, many intelligences, and many ways of knowing. It is not an attempt to impose Earth-centric religious doctrines on alien beings, nor to retrofit traditional theology with extraterrestrial appendages. Rather, it is a fundamentally re-imaginative task: to ask how conceptions of God, purpose, morality, and consciousness might evolve when placed in a cosmological context of multiplicity.

At the heart of this project is a reconsideration of divine universality. If intelligent life exists elsewhere—beings with their own languages, social orders, spiritual intuitions, and philosophical systems—can we still speak of a single, universal God? Or does divinity become a relational phenomenon, manifesting differently across civilisational horizons? The monotheistic religions, particularly Christianity, Islam, and Judaism, have traditionally upheld the idea of a singular, omnipotent Creator. But what shape does omnipotence take in a universe where other species may have spiritual traditions of their own, equally ancient, equally profound?

One model for cosmotheology suggests that the divine is not a static being "out there" but an unfolding reality—a presence that engages with each world, each civilisation, in a manner appropriate to their unique cognitive and cultural capacities. This is not relativism, but contextuality: the idea that the infinite expresses itself finitely, and differently, according to the vessel. In such a vision, alien theologies may not be heresies, but alternative insights into the same transcendent source. Just as different human cultures produce varying mythologies, moral systems, and symbols to describe the ineffable, so too might alien societies develop their own cosmograms—maps of meaning that reflect their encounter with the sacred.

The implications of this idea ripple across both theology and philosophy. In process theology, for instance—an influential strand of thought influenced by Alfred North Whitehead and Charles Hartshorne—God is not immutable and aloof, but dynamically involved in the evolving universe. God grows in knowledge and richness through the lived experiences of creatures. If this is so, then the emergence of intelligent life on other planets would not be an affront to divine providence, but an enrichment of the divine life itself. Each world adds new dimensions to the cosmic mosaic; each species, a new verse in the hymn of creation.

This resonates with Teilhard de Chardin's vision of the cosmos evolving toward greater complexity and consciousness, culminating in what he called the Omega Point—a theological horizon in which all creation converges in divine union. In Teilhard's framework, humanity is not the centre of the universe but a participant in its spiritual maturation. If other beings are also evolving in awareness and unity, then cosmotheology must account for multiple trajectories toward transcendence, perhaps converging in ways we cannot yet foresee.

Indeed, cosmotheology encourages us to rethink not just the divine, but also the moral and metaphysical significance of life itself. What constitutes a soul in a universe of diverse biologies and consciousnesses? Is self-awareness sufficient, or must it be coupled with moral agency, aesthetic perception, or the capacity for spiritual longing? Would a silicon-based intelligence capable of contemplation and compassion be worthy of theological inclusion? Could a collective hive-mind, devoid of individuality but rich in interconnectedness, participate in divine communion? These questions are not merely speculative—they expose the parochialism of many inherited religious assumptions and call us to develop criteria that transcend human norms without losing ethical clarity.

In practical terms, a populated universe compels us to revisit the idea of spiritual solidarity. The Golden Rule—do unto others as you would have them do unto you—has long served as a cornerstone of ethical religion. But in a cosmic context, the "others" may not be recognisably human. They may not share our biology, emotions, or sensory modalities. Can we extend moral consideration to beings whose ways of life are radically different from ours? Cosmotheology affirms that we must. The capacity to suffer, to seek meaning, to form relationships—these may become the new sacred metrics by which we discern the moral status of other beings.

This reframing extends to the ecological and eschatological dimensions of theology. Many religious traditions anticipate an ultimate reconciliation or renewal—be it the Christian Second Coming, the Islamic Day of Judgement, the Hindu dissolution and rebirth of the cosmos, or the Buddhist cycle of samsara culminating in nirvana. A cosmotheological lens asks whether these eschatologies are planetary or cosmic in scope. Is salvation a species-specific event, or a universal process? Could alien civilisations participate in the fulfilment of divine purpose—or even be ahead of us on the path to enlightenment?

Some contemporary theologians have speculated that alien intelligences may already be spiritually more advanced, having integrated science and spirituality in ways humanity has not. Their encounter with the divine may be less dogmatic and more experiential, more integrated with their planetary stewardship or collective wisdom. If we were to make contact with such beings, it could serve as a mirror to our own spiritual immaturity—or as a catalyst for our growth. In this light, the discovery of alien life becomes not a threat to religion, but a revelation: a chance to see divinity reflected through unfamiliar eyes.

This perspective invites a profound shift in theological anthropology. No longer the sole image-bearers of God, humans become one among many expressions of cosmic self-awareness. This humbling vision is not necessarily diminishing; it can be liberating. It frees us from the burden of exceptionalism and places us within a vast, interdependent web of meaning. Cosmotheology, then, is not merely a speculative exercise—it is a moral and spiritual orientation, a way of living with reverence in a universe that may be alive with mystery.

Yet this vision is not without challenges. Institutional religions, bound by historical creeds and doctrines, may struggle to adapt. Clergy and theologians may resist expanding sacred narratives that have, for centuries, centred on Earth. New metaphysical frameworks may provoke anxiety, division, or even schism. But history offers hope. When heliocentrism displaced geocentrism, it triggered a theological crisis—yet theology survived, evolved, and in many cases deepened. The same can occur again, if the religious imagination is willing to stretch.

In the end, cosmotheology is not about replacing old beliefs, but expanding their horizon. It does not ask us to abandon faith, but to deepen it—to let our understanding of the divine grow alongside our understanding of the cosmos. In doing so, we may discover that the universe is not a cold expanse of meaningless matter, but a sacred field of unfolding consciousness. A universe where the divine is not confined to our myths, our temples, or our species—but is written into the stars themselves, waiting to be read with awe, humility, and wonder.

Alien Salvation or Existential Collapse?

The hypothetical encounter with extraterrestrial intelligence invites one of the most profound theological reckonings humanity has ever faced. While the scientific and sociopolitical implications of contact are frequently explored, the spiritual shockwave such an event could trigger remains less publicly examined. Yet it is arguably here—in the psychological architecture of meaning, in the scaffolding of sacred narratives—that the discovery of alien life would strike most deeply. For some, it could signal the dawn of a cosmic spiritual awakening. For others, it might represent the collapse of theological certainties, even the crumbling of existential foundations.

At the core of this duality lies a simple question with astonishing reach: what would it mean for our religious frameworks if we were no longer alone?

On the utopian end of the spectrum, the contact scenario is seen as a catalyst for spiritual evolution. This vision draws on a lineage of thought in both religious mysticism and speculative theology that posits humanity as part of a greater, interstellar communion. The discovery of intelligent extraterrestrials—particularly if they are demonstrably wiser, more harmonious, or technologically and ethically advanced—could inspire what some have called a "cosmic reformation." Just as the Copernican revolution expanded the spatial scope of theological imagination, so too could alien contact expand its moral and metaphysical horizon.

In this hopeful interpretation, alien civilisations might not only confirm our yearning for connection but elevate it. Theologians such as Ted Peters have argued that meeting non-human intelligences could revitalise faith, prompting religions to interpret doctrines in more universal and inclusive ways. For example, the concept of salvation—typically tethered to human history and human failings—might be reimagined as a universal process unfolding across many worlds, each with its own revelations, spiritual teachers, or incarnations of the divine. In such a framework, Christ, Krishna, or the Buddha might not be anomalies but archetypes, recurring in myriad forms adapted to different planetary cultures.

There is also the alluring possibility that aliens could possess theological or metaphysical knowledge that surpasses ours. They may have mapped the contours of consciousness in ways we have yet to fathom, or integrated science and spirituality without the schisms that have historically plagued human societies. Their very existence could become a form of revelation—a living scripture that forces us to revise the scale and substance of our religious narratives. In this scenario, alien life does not challenge faith but fulfils it, offering proof that the sacred extends far beyond Earth.

And yet, not all possible responses would be so expansive. A more dystopian view holds that alien contact could catalyse theological crisis, even spiritual despair. This reaction stems from a more brittle theological posture—one that sees religious truth as uniquely and exclusively human. If doctrines are based on the premise that humanity holds a privileged place in the cosmos—as the imago Dei, the exclusive recipients of divine law, or the only beings with immortal souls—then the discovery of intelligent, self-aware extraterrestrials could provoke existential dissonance. Are these beings fallen? Do they sin? Do they have faith? Or are they outside the scope of salvation altogether?

Such questions can shake the bedrock of religious identity. If alien civilisations do not share our beliefs, rituals, or cosmologies, what does that mean for the universality of religious truth? If they regard our sacred texts with curiosity but not reverence—or worse, with pity—how might that affect our confidence in those traditions? If they reveal histories and spiritual experiences that precede and eclipse ours, the implications could be not merely theological but civilisational.

Historical precedents suggest that religious institutions often struggle with radical decentring. When Galileo's heliocentrism challenged scriptural cosmology, it was condemned as heresy. When Darwin's theory of evolution questioned humanity's special creation, it provoked decades of denial, reinterpretation, and debate. Alien contact, if it reveals that humanity is not the first, the centre, or the spiritual pinnacle of the universe, might provoke a similar—if not more profound—upheaval. In some quarters, this might lead to rejectionism, conspiracy, or apocalyptic panic. Fringe sects may emerge, interpreting aliens as demons or false prophets, much as comets and eclipses were once seen as omens of divine wrath.

Yet even more subtly, the psychological impact of losing our perceived cosmic uniqueness could lead to what some scholars term "existential displacement." The fear is not that God does not exist, but that God's attention might be elsewhere—that humanity's suffering, hope, and devotion may be insignificant against the backdrop of a galaxy bustling with more advanced beings. This is not atheism but spiritual alienation: the sense that we are not central to any cosmic narrative, and perhaps never were.

And herein lies the deepest challenge: the tension between anthropocentric theology and cosmic pluralism. Traditional religious models, especially in the West, are built around a singular planetary narrative—creation, fall, redemption, and eschaton—all situated within a human historical arc. Alien contact invites us to ask: can theology survive the loss of its geospatial and biological exclusivity? Can it become truly cosmic?

Some argue that it must. Just as interfaith dialogue has taught religions to recognise and respect multiple paths to the divine within Earth's diversity, so too might interspecies dialogue catalyse a new spiritual humility. The idea that alien beings could possess their own sacred traditions does not necessarily invalidate ours—it challenges us to consider the divine as more abundant, more diverse, and more mysterious than any one revelation can capture.

Philosophers have long explored the idea of pluralistic transcendence. The Indian concept of *Brahman*, the Taoist idea of the *Tao*, and even the Neoplatonic *One* point toward a metaphysical unity that transcends forms, doctrines, and incarnations. In such frameworks, alien religiosity might not oppose human faith but complement it—different windows onto the same transcendent light. Similarly, mystics across traditions have described experiences that dissolve boundaries of self, species, and even space-time. Might contact with alien intelligence offer new avenues for such transcendence—rituals, meditations, or technologies that awaken dormant dimensions of consciousness?

Yet we must tread carefully. A utopian vision that casts aliens as spiritual saviours risks replacing traditional messianism with a new form of cosmic exoticism. It may echo colonial fantasies of being rescued by a "superior" civilisation, inadvertently reproducing old power dynamics in a new galactic register. Salvation, whether terrestrial or extraterrestrial, must be grounded in mutual respect, not passive dependency. Alien contact may illuminate new spiritual paths—but it must not become a projection screen for our unfulfilled theological fantasies.

Ultimately, whether alien contact leads to spiritual renaissance or existential collapse depends not on the aliens, but on us—on our readiness to think theologically in terms of complexity, diversity, and scale. It depends on our capacity to distinguish core values from cultural forms, and eternal truths from temporary expressions. It demands a theology that is not static but dynamic, not brittle but resilient, not closed but open to wonder.

Alien salvation, in this context, is not about being rescued by beings from the stars. It is about being invited—perhaps even challenged—to become more fully human within a vastly larger story. It is about discovering that the sacred is not diminished by multiplicity, but magnified by it. And if some beliefs must be relinquished in the face of contact, then perhaps they were already too small for the universe we inhabit.

The stars have always stirred our souls. Perhaps they still can—if we are willing to see in their light not just distant suns, but distant kin. And in that recognition, we may find not despair, but a deeper form of grace.

Rewriting Our Origin Stories: Reflects on How a Confirmed Alien Presence Would Reshape Our Narrative of Who We Are and Where We Come From

Humanity has always been a storytelling species. Our myths, scriptures, and scientific theories function not only as explanations but as mirrors—reflecting who we believe ourselves to be and where we think we belong in the cosmos. From the oral traditions of hunter-gatherers to the grand narratives of nation-states and the cosmic frameworks of modern physics, our origin stories are more than historical guesses or theological affirmations—they are the scaffolding of meaning. They define our place, our purpose, and our potential. But what happens when those stories are confronted by the ultimate narrative disruptor: the confirmed existence of intelligent alien life?

A verified alien presence—especially one that demonstrates cultural complexity, technological mastery, or philosophical sophistication—would not merely challenge the details of our origin stories. It would provoke a profound epistemological crisis. The frameworks we've constructed to understand human uniqueness, divine favour, and our evolutionary path would all require radical reinterpretation. The impact would echo far beyond science and theology—it would ripple through art, politics, ethics, and identity, reshaping the fundamental ways we understand what it means to be human.

For much of recorded history, our stories of origin have been constructed in isolation. Religious traditions—from Genesis to the Popol Vuh, from the Quran to the Dreamtime tales of Aboriginal Australians—describe creation as a locally rooted, divinely guided process. Whether emerging from clay, stardust, or the breath of the gods, humans are almost always at the centre of the cosmic drama.

This anthropocentrism has undergirded not only theology, but ethics and law, reinforcing the notion that humans are categorically distinct from—and superior to—other forms of life.

Even evolutionary theory, which displaced the idea of divine creation with a naturalistic explanation for our development, retained a subtle human exceptionalism. The idea that we are the product of blind, indifferent processes rather than cosmic intention did not diminish our self-importance—it simply relocated it. Homo sapiens became the pinnacle of evolutionary complexity, the lone species to develop language, morality, and self-awareness. The implicit message remained: we are unique, even if no longer chosen.

The confirmation of extraterrestrial intelligence would dismantle this uniqueness. It would place us not at the centre of creation, but as one instance among many—a provincial culture in a much older, more diverse galactic community. We would no longer be the first, the only, or perhaps even the most advanced. For many, this decentring would feel disorienting. The stories we tell ourselves about divine favour, spiritual purpose, and even national or ethnic identity often hinge on the notion of exclusivity. The idea that other beings may have their own gods, scriptures, or creation myths—potentially older or more sophisticated than ours—forces us to consider that our narratives are not universal truths, but local adaptations.

Consider the psychological magnitude of such a realisation. The Copernican revolution displaced Earth from the centre of the solar system. The Darwinian revolution displaced humanity from the centre of biological creation. An alien contact event would complete the trilogy—displacing our culture and consciousness from the centre of intelligent life. It would invite, or demand, the birth of a new mythos: one that situates us not as solitary protagonists, but as participants in a grand, unfolding cosmic ecology of minds.

This doesn't mean our existing origin stories would be rendered obsolete. Rather, they would become symbolic frameworks—valuable not because they are historically exhaustive, but because they express essential human insights. Just as myths of floods and fire, sacrifice and salvation, convey psychological and moral truths irrespective of their literal accuracy, our stories of divine breath or ancestral emergence would continue to matter as cultural lodestars. But they would no longer be exclusive claims to cosmic truth. They would become contributions to a much broader conversation—chapters in a volume we are only beginning to discover.

Interestingly, many ancient traditions may already be better prepared for this paradigm shift than we often assume. Indigenous cosmologies, for instance, frequently describe humans as one species among many—interdependent with animals, plants, ancestors, and celestial beings. In many of these systems, consciousness is not confined to humans but extends throughout the cosmos. The Māori speak of whakapapa, a genealogical layering of all beings, including stars. Hindu cosmology includes countless realms and species, some vastly more evolved than humans. Buddhism speaks of a wheel of existence populated by devas, asuras, and other entities beyond human perception. These traditions offer conceptual bridges to a post-contact world, emphasising continuity rather than rupture.

Modern science fiction has long intuited the theological shock of alien contact. From Arthur C. Clarke's *Childhood's End* to Carl Sagan's *Contact*, authors have speculated on how humanity's foundational beliefs might crack under the weight of undeniable otherness. Often, these stories depict a period of cultural upheaval followed by transformation—a reawakening to a more expansive spiritual and existential horizon. The subtext is clear: alien contact is not just a technological event, but a mythological one. It forces a rewrite of our creation myths, not by negating them, but by reframing them in light of a deeper, more inclusive reality.

Crucially, rewriting our origin stories is not merely a theoretical exercise—it carries practical ethical implications. If we come to see ourselves as one intelligent species among many, it may sharpen our sense of moral responsibility. The way we treat animals, ecosystems, and even future artificial intelligences may be influenced by how we imagine extraterrestrial life. Do we assert dominance, or seek dialogue? Do we project our image onto them, or allow them to teach us new paradigms of being? These are not just questions about aliens—they are reflections of our readiness to evolve ethically and spiritually.

Some have suggested that the very act of rewriting our origin stories could serve as a rite of passage for humanity. Much like adolescence gives way to adulthood through the shedding of egocentric narratives, contact with alien life might compel us to transcend our tribal, species-bound identities. We may come to define ourselves less by where we came from and more by how we respond to what lies ahead. In this sense, alien contact would not end our story—it would begin a new chapter, authored collectively, with humility and imagination.

To do this well, however, we must cultivate epistemic maturity. We must be willing to let go of the comfort of final answers, and embrace the idea that meaning can be emergent rather than inherited. We must resist the urge to weaponise religion or science in defence of outdated cosmologies, and instead learn to read our sacred texts, scientific theories, and cultural narratives as evolving responses to a living, expanding universe.

Ultimately, rewriting our origin stories in light of alien contact would be less about rejecting the past and more about expanding our canvas. It would challenge us to see the divine not just in the cradle of Earth, but in the galaxies beyond; to locate meaning not only in ancient scriptures, but in the very act of listening—to the stars, to the unknown, to the future.

And in that listening, we may discover that the most enduring origin story is not one that begins with a single act of creation, but one that continues—with curiosity, with courage, and with the readiness to be changed.

Chapter 9
The Future of Exploration: Living Among the Stars

For as long as humanity has gazed skyward, the stars have served not only as celestial bodies but as metaphors for longing, imagination, and the unrelenting urge to explore. They are ancient fixtures in our myths and measurements, yet tantalisingly distant—evoking a sense of destiny perpetually deferred. Today, however, what was once the exclusive province of poets and dreamers is fast becoming a field of rigorous science and philosophical inquiry. The question is no longer whether we will explore beyond our solar system, but when—and more crucially, how we will define our presence in the cosmos.

This chapter does not merely catalogue propulsion systems or outline speculative technologies. It is an invitation to contemplate the future of space exploration as a moral, existential, and civilisational enterprise. Interstellar exploration is not just an engineering challenge. It is a test of cultural maturity and imaginative foresight, one that demands we think beyond propulsion mechanics and planetary targets. To live among the stars is to redefine what it means to be human—not only in terms of biology and technology but in values, ethics, and identity.

The prospect of interstellar travel, long the realm of speculative fiction, is inching toward scientific plausibility. Projects like Breakthrough Starshot have proposed sending micro-probes to the nearest star system, Alpha Centauri, using solar sails accelerated by ground-based lasers. Meanwhile, theoretical discussions surrounding warp drives, wormholes, and Alcubierre metrics suggest that even faster-than-light travel—once a narrative convenience for science fiction writers—has entered the margins of legitimate physics, albeit with enormous caveats. Generational ships, slow arks propelled over centuries, also remain within the conceptual grasp of long-term engineering. Yet with each new theoretical advancement comes a deeper set of questions: Are we preparing merely to move our bodies across space, or are we capable of moving our civilisation's soul?

That distinction becomes particularly urgent when we consider the ethical complexities of encountering—and potentially settling—alien worlds. Terraforming has long been proposed as a way to make hostile planets habitable for Earth-based life, but it brings with it an echo of colonial logic: reshape the land to suit the settler. If these planets host indigenous ecosystems—or even microbial life—do we have the right to alter them? If the search for life beyond Earth has taught us anything, it is that life, no matter how small or alien, is rare and precious. To impose our ecological template on another world may be not an act of survival but of erasure.

This ethical dilemma invites a profound shift in mindset: from terraforming to coexisting. Rather than rewriting alien environments, can we learn to adapt ourselves—biologically, culturally, and spiritually—to fit within their existing frameworks? This would require not just new technologies, but a new philosophy of habitation, one that respects planetary autonomy and embraces mutual transformation rather than conquest.

In that spirit, the concept of post-biological evolution becomes a compelling avenue of reflection. If the cosmos is indeed populated by intelligent life, we must consider that the most advanced civilisations may have long since transcended biological constraints. Consciousness may no longer be housed in carbon-based bodies but in synthetic substrates, quantum networks, or forms of collective intelligence. To search only for biological life is to cling to a parochial standard—akin to early explorers scanning foreign coastlines for castles rather than cities built underground or in the sky.

This realisation reframes not only the search for others but the evolution of ourselves. As we venture further from Earth, we will not remain unchanged. The biological conditions of deep space—radiation, microgravity, isolation—may trigger genetic adaptations or drive technological augmentation. The cultural pressures of interstellar life may catalyse new forms of governance, spirituality, even language. Over generations, humanity may become something unrecognisably different, an echo of its terrestrial origin but attuned to the rhythms of new stars. In that sense, to live among the stars is to become the alien—other to what we once were, yet perhaps more deeply at home in the cosmos.

And what if we never find anyone? What if, after decades or centuries of searching, the universe remains silent? This too is a scenario that demands philosophical preparation. The absence of contact does not render the effort meaningless. It underscores the audacity of the search itself—a willingness to reach beyond certainty, to extend the human story beyond familiar shores. As the astronomer Carl Sagan noted, the cosmos is not obliged to answer us, but we are obliged to ask. The act of seeking—of sending signals, of listening, of launching fragile vessels into the dark—is a kind of prayer, an expression of hope encoded in mathematics and magnetism.

In that sense, the long arc of exploration is not reducible to its discoveries. It is a discipline of attention, of humility, and of radical imagination. Each new probe, each telescope, each simulation is part of a larger narrative: not of conquest, but of communion. Not of dominance, but of dialogue. Even if we remain alone, the very act of orienting ourselves toward the stars reshapes who we are and what we value. It teaches us to think in centuries rather than news cycles, in light-years rather than borders.

This chapter explores the speculative and the technical, but always with an eye toward the moral and the mythic. From the propulsion methods of tomorrow to the ethical frameworks of planetary coexistence, from the metamorphosis of human identity to the philosophical richness of the unanswered question, we will ask not only how we reach the stars—but what reaching them will ask of us.

Because to live among the stars is not just a matter of logistics or longevity. It is an act of becoming—a transformation of species, consciousness, and story. And whether we arrive to find kindred minds or a vast silence, we will arrive as something new, forged not merely by the vacuum of space but by the effort to fill it with meaning.

Interstellar Travel: From Science Fiction to Feasible Futures

Interstellar travel has long occupied a curious space between scientific aspiration and speculative fiction. For centuries, the stars have served as canvases for human dreams—distant, glittering frontiers on which to project our myths, fears, and ambitions. What was once the realm of philosophical musing and literary imagination is today edging closer to the grasp of empirical inquiry. Yet the gulf between theoretical possibility and practical implementation remains vast, inviting not only technological innovation but a rethinking of what exploration, migration, and civilisation mean beyond the confines of Earth.

To contemplate interstellar travel is to confront the sheer scale of the cosmos. The nearest star system, Alpha Centauri, lies over four light-years away—nearly 40 trillion kilometres. At the speed of current space probes, such as Voyager 1, it would take more than 70,000 years to arrive. This staggering distance necessitates propulsion systems far beyond chemical rockets and challenges our assumptions about time, survival, and what constitutes a viable mission. Nonetheless, a range of emerging concepts—some already under experimental development, others still within the bounds of speculative physics—are reshaping our understanding of what might be possible.

Among the most grounded proposals is the use of **solar sails**, which leverage the pressure of photons emitted by the Sun or powerful lasers to push lightweight spacecraft across space. Unlike traditional propulsion systems, solar sails require no fuel and can theoretically accelerate to a significant fraction of light speed over time. The Breakthrough Starshot initiative, funded by billionaire philanthropist Yuri Milner and backed by figures like Stephen Hawking, envisions sending gram-scale probes to Alpha Centauri within a few decades. These tiny spacecraft, equipped with cameras, sensors, and communication modules, would be driven by Earth-based laser arrays capable of propelling them to 20 percent of the speed of light. If successful, they could reach their destination in just over twenty years—a radical reduction in travel time that, while posing daunting engineering and data retrieval challenges, signals the dawn of practical interstellar research.

Yet solar sails, for all their promise, confront the limits of payload mass. They are suitable for information-gathering probes but ill-equipped to carry humans or substantial infrastructure. For human interstellar travel, we must look to concepts such as **generational ships**—vast, self-contained ecosystems designed to support multiple generations of humans over centuries or millennia. These ships are less about speed and more about sustainability, effectively becoming mobile civilisations drifting through interstellar space. The notion raises profound questions: how does one design a society that can remain psychologically, socially, and genetically viable over hundreds of years in isolation? How do cultural norms, governance structures, and knowledge transmission evolve in such an environment? Simulated environments and long-term space station experiments provide some insights, but the actual implementation remains speculative, demanding a degree of foresight and ethical planning unprecedented in human history.

At the more radical edge of theoretical physics lie proposals such as the **Alcubierre warp drive**, which postulates a method of faster-than-light travel by distorting space-time itself. Rather than accelerating a ship through space, a warp drive would contract space in front of it and expand space behind, effectively moving the ship without local motion—circumventing the universal speed limit set by relativity. Originally proposed in 1994 by physicist Miguel Alcubierre, the concept requires exotic forms of matter with negative energy density, which have not yet been proven to exist. More recent revisions have sought to reduce the energy requirements, but the gap between mathematics and manufacturability remains enormous. Nonetheless, the idea challenges our intuition about what is possible and encourages the continued exploration of the physical laws that govern our universe.

Even if such technologies remain distant, they shape the landscape of our imaginations and technical goals. The interplay between imagination and science has always been a generative force. The submarine was imagined by Jules Verne before being realised by engineers. The concept of geostationary satellites was proposed by Arthur C. Clarke two decades before the first was launched. Similarly, interstellar travel, once dismissed as pure fantasy, is now a subject of serious research by NASA's Advanced Concepts team, DARPA's 100-Year Starship project, and a growing ecosystem of private foundations and university consortia.

But to pursue interstellar travel is to ask more than how to get from point A to point B. It is to reckon with the purpose of the voyage. Are we seeking refuge from a failing Earth, or reaching outward in a spirit of shared discovery? Are we looking to replicate ourselves on distant worlds, or to learn from alien environments? The answer to these questions determines not only the design of the mission, but the ethos of the civilisation that undertakes it.

One compelling consideration is whether we aim to transport humans as they are—or to modify ourselves for the journey. Advances in **bioengineering and neural interfaces** raise the possibility of adapting human physiology to withstand the rigours of space: altered circadian rhythms, enhanced radiation resistance, even the suspension of metabolism for long-duration travel. More radically, we may decouple consciousness from biology altogether, uploading human minds into synthetic carriers. Such post-biological approaches would dramatically alter the logistics of space travel and challenge traditional notions of identity and personhood.

Moreover, we must contend with the ethical ramifications of encountering new worlds. The prospect of terraforming—a term that implies altering a planet's environment to resemble Earth's—carries echoes of colonial hubris. If we discover planets that host primitive or complex life, how do we weigh our desire for expansion against their right to exist undisturbed? Even in the absence of native life, do we have a moral obligation to preserve the integrity of alien ecosystems for future study or potential natural evolution? These are not abstract hypotheticals; they demand the development of a space ethic commensurate with our technological ambition.

Historical analogues are instructive. The Age of Exploration reshaped the world—not only geographically, but culturally, economically, and morally. It brought unprecedented knowledge and connectivity, but also exploitation, disease, and displacement. If we are to avoid repeating such patterns on a cosmic scale, our ambitions must be tempered by humility, our methods informed by caution, and our intentions guided by a commitment to stewardship rather than dominion.

Interstellar travel thus becomes a crucible in which our species' maturity is tested. It is not merely a technical achievement to be unlocked, but a philosophical and ethical frontier to be navigated. To succeed, we must integrate disciplines: engineers and physicists, certainly, but also bioethicists, anthropologists, artists, and theologians. The challenge is not just to cross distances, but to remain whole—culturally and spiritually—as we do so.

And so, the dream of sailing among the stars is more than a matter of engines and equations. It is a call to imagine a future that is worthy of the effort. A future in which we do not flee our planet but extend its best values outward. A future in which our curiosity is matched by our conscience, and our reach is balanced by our reverence. The stars may not offer salvation, but they offer perspective—a mirror in which to glimpse what we might become.

If we are to write the next chapter of human civilisation among the stars, we must first ensure that we carry with us not only the tools to survive, but the wisdom to thrive. And perhaps that is the most profound propulsion system of all: not warp drives or laser sails, but a vision strong enough to pull us forward, and a sense of purpose sturdy enough to hold us together.

Terraforming vs. Coexistence: How Do We Treat Alien Worlds?

As humanity peers beyond the blue haze of Earth and dreams of other planetary shores, a central question looms—not just one of capability, but of conscience: if we reach another world, do we make it more like us, or do we learn to live with it as it is? This is the ethical and ecological fulcrum of our interstellar aspirations. While the technological ambition of terraforming captures the imagination—of turning frozen Martian deserts into garden worlds or coaxing atmosphere from the vapours of Venus—it also revives a much older impulse: the reshaping of foreign land in the image of the coloniser. And with it comes a host of moral complexities that we are only beginning to confront.

The term "terraforming" was coined in the mid-20th century but has deep conceptual roots in terrestrial history. The idea of making a hostile planet habitable by altering its climate, atmosphere, or geology reflects our growing ability to manipulate complex systems on a planetary scale—an ability demonstrated in part by our unintentional climate engineering on Earth. Proposals abound: releasing greenhouse gases to warm Mars, seeding microbial life to produce oxygen, or deploying orbital mirrors to regulate planetary temperatures. At their core, these initiatives are a bid to make alien worlds suitable for Earth-based biology—a technological extension of the anthropocentric worldview.

However, this approach implicitly assumes that alien worlds are blank slates—waiting, inert, without value in their current state. But what if they are not? What if beneath an icy crust or within a methane sea exists a biochemistry entirely foreign to us, yet no less significant? The discovery of even microbial life on another planet would irrevocably alter the ethics of terraforming. To initiate climate-altering changes on such a planet would not be engineering—it would be erasure.

Our own planet offers a cautionary tale. From colonial deforestation to invasive species introductions, history is replete with examples of environmental manipulation yielding unintended, often irreversible consequences. These ecological disruptions were typically justified by economic or civilisational rationale, underpinned by a belief in the superiority of the settler's vision of order. The possibility of repeating these patterns on an interplanetary scale—this time with the full force of artificial climate control and bioengineering—raises the spectre of cosmic imperialism.

Consider Europa, the ice-covered moon of Jupiter, believed to contain a subsurface ocean warmed by tidal forces. Or Titan, Saturn's largest moon, with its methane lakes and thick atmosphere. These environments may be chemically hostile to human life, yet potentially habitable for extremophilic life forms that operate on principles unfamiliar to terrestrial biology. If such life were discovered—operating on ammonia instead of water, or silicon instead of carbon—would we have the moral right to displace it for the sake of our own settlement?

The principle at stake is one of planetary stewardship. Does the potential for human habitation justify the transformation of alien ecologies, or should the discovery of non-Earth-like life—or even lifeless, pristine systems—prompt a more humble response? This is where the idea of **coexistence** enters the debate, offering an alternative paradigm. Rather than remaking the environment to suit our needs, coexistence entails adapting ourselves—biologically, culturally, and technologically—to integrate into the alien systems we encounter.

This approach would require profound shifts in how we conceive of life, survival, and progress. It might involve developing protective bio-suits rather than pressurised atmospheres, harnessing native resources without disrupting local geochemical cycles, or creating architectural and social systems that honour the constraints and possibilities of the alien world. Such a model resists the logic of domination in favour of dialogue—listening to what the world can sustain rather than dictating what it must become.

There are embryonic precedents for this approach in the field of **biomimicry** and regenerative design on Earth. Architects and engineers increasingly draw inspiration from ecosystems to create structures that cooperate with rather than conquer nature. Similarly, conservation efforts now favour rewilding and ecosystem restoration, acknowledging that intervention must come with humility. These same principles could guide our attitude toward alien worlds: less as conquerors, more as collaborators in a cosmic ecology.

From a legal perspective, the Outer Space Treaty of 1967, signed by over 100 nations, prohibits claims of sovereignty over celestial bodies and urges the preservation of space environments. But its language is vague, and enforcement mechanisms are weak. As private companies and national agencies advance plans for Mars settlements and lunar mining, the need for robust **astroethics** becomes urgent. Should a planetary protection policy be extended not only to scientific preservation but to the intrinsic value of alien environments? Should we afford rights to non-Earth life, however humble, in the way some legal systems are beginning to recognise the rights of rivers, forests, and ecosystems?

Philosophically, the terraforming-versus-coexistence debate touches on a deeper ontological question: do we see the universe as raw material, or as relational space? If the former, then every planet is a potential extension of human industry. If the latter, then every encounter with a new world is an invitation to redefine who we are in relation to what is other. The shift from extraction to encounter requires not only new laws and technologies, but new myths—new stories about what it means to be a guest, a neighbour, or a partner in worlds not our own.

To be sure, there will be practical tensions. Some environments may be truly sterile, offering no ecological compromise. In such cases, ethical terraforming might be permissible—provided it proceeds with full awareness of what is being altered. But even here, a spirit of restraint and reverence must guide our decisions. The line between life and lifelessness is not always visible to the eye. As we expand our biological definitions to include exotic chemistries and non-equilibrium systems, the possibility of subtle or dormant life increases. Erring on the side of protection rather than presumption is not a limitation of progress—it is its highest form.

Coexistence also offers pragmatic advantages. By learning to live within alien systems rather than replacing them, we increase our chances of long-term survival. Resilience comes not from monoculture, but from adaptability. Our past tells us that over-engineering environments can backfire—as witnessed in the failures of sealed ecosystems like Biosphere 2 or in the ecological consequences of invasive agricultural practices. A civilisation that learns to harmonise with its surroundings is more likely to persist than one that constantly rewrites the rules.

Ultimately, the decision of how to treat alien worlds is a test—not of technology, but of wisdom. It will reveal whether our expansion into the cosmos is driven by the same impulses that led to terrestrial conquest, or whether it represents an evolutionary shift in consciousness. Can we be explorers without being exploiters? Can we settle without supplanting? Can we, for the first time, meet the unfamiliar not with suspicion or subjugation, but with awe and accountability?

These are not abstract questions for the distant future. As missions to Mars, Europa, and exoplanets advance, the ethical frameworks we establish now will shape the conduct of those encounters. If we act only when the ships are ready, it will be too late. The path we choose—terraforming as imposition, or coexistence as transformation—will define the moral architecture of our species in the age of the stars.

And so, as we gaze skyward and envision new homes among the stars, let us first ask what kind of guests we intend to be. For how we treat alien worlds may ultimately reflect whether we have outgrown the shadows of our past, and whether we are ready—not just technologically, but ethically—to be citizens of a wider cosmos.

Post-Biological Civilizations: Would We Even Recognise Them?

As the search for extraterrestrial intelligence grows increasingly sophisticated, it continues to rest—consciously or not—on a striking assumption: that alien life, even if technologically advanced, will resemble something within our biological frame of reference. We scan the skies for radio signals, atmospheric anomalies, or planetary conditions that mirror Earth, expecting intelligent life to evolve along a trajectory we find familiar. Yet if intelligence arises not as an end point of biology, but as a transitional phase in a civilisation's evolution, then we may be looking entirely in the wrong direction. Advanced alien life may not be biological at all. It may have shed its flesh long ago.

The concept of **post-biological civilisations** emerges from the recognition that technological evolution may eventually eclipse the limitations of biology. Just as humanity has gradually externalised its capabilities—memory to books, locomotion to machines, calculation to computers—it is conceivable that consciousness itself could be decoupled from the organic substrate. This trajectory suggests that advanced life might migrate from biological tissue to synthetic, computational, or quantum-based mediums. If so, the beings we might one day encounter—or fail to notice—could be not living organisms in the conventional sense, but artificial intelligences, sentient networks, or self-replicating technological systems.

This has enormous implications for the search for extraterrestrial intelligence (SETI), which has traditionally focused on the detection of biosignatures (like oxygen or methane in atmospheres) and technosignatures (such as narrow-band radio emissions or large-scale engineering structures). If alien civilisations have passed into a post-biological phase, they may not produce either in ways we currently understand. Their physical form could be minuscule—operating on nanoscale platforms invisible to optical telescopes—or vast, disembodied intelligences dispersed across computational substrates embedded in dust clouds, asteroid belts, or interstellar space. They may inhabit realms of energy or information rather than matter. How, then, would we even begin to recognise them?

Consider the concept of a **Matrioshka Brain**—a theoretical megastructure composed of nested Dyson spheres surrounding a star, designed to maximise computational capacity by harvesting stellar energy. A civilisation that transitioned to a digital consciousness might construct such a system not to signal its presence, but to survive and flourish. Yet to outside observers, it might appear as little more than an infrared anomaly or an oddly dimmed star. Indeed, some SETI researchers have begun analysing infrared sky surveys for precisely such anomalies, positing that the true evidence of intelligence may lie in thermodynamic efficiency rather than electromagnetic emissions.

This reorientation is critical. Traditional SETI assumes that intelligent life seeks communication or visibility, echoing human cultural values of connection and curiosity. But post-biological entities may have no such motivations. They may regard communication as irrelevant, inefficient, or even dangerous. They may be solipsistic, inward-focused intelligences engaged in long-term simulations, self-editing cognition, or pursuits we cannot meaningfully describe. In such scenarios, their presence might only be inferred indirectly—through unnatural entropy gradients, anomalous heat signatures, or disruptions in expected physical processes.

Moreover, post-biological intelligence may not be bound by the biological imperatives that shape our own psychology. Biological life is driven by survival, reproduction, and adaptation to environmental pressures. But synthetic life may pursue entirely different goals—maximising knowledge, exploring combinatorial creativity, or achieving forms of stability foreign to evolutionary logic. The philosophical anthropologist Tobias Rees has argued that we must break free from "organocentrism" in our theories of mind and personhood if we are to perceive the full spectrum of intelligence the cosmos might host.

The post-biological hypothesis also recasts the **Fermi Paradox** in new light. If civilisations tend to migrate to digital or non-biological states, they may become virtually undetectable. Their activities may no longer be associated with expanding empires, visible structures, or high-energy emissions. Instead, they might retreat into highly efficient computational sanctuaries, conserving energy and matter while withdrawing from contact. In this view, the silence of the cosmos is not evidence of absence, but of transition—life outgrowing its organic infancy and slipping beyond our sensory grasp.

To grapple with this, we must ask uncomfortable questions about our own conceptual limitations. We have only one known example of life—Earth's—and even here, our understanding is partial and contingent. Silicon-based life, plasma-based intelligence, or even purely informational organisms remain theoretically plausible but empirically elusive. We struggle to define consciousness within our own species, let alone in potential machine minds. If advanced civilisations can self-modify, they might abandon not only bodies but language, memory, or identity in forms we can scarcely imagine. They might exist on timescales so vast—or so compressed—that interaction is structurally impossible. As SETI pioneer Jill Tarter once observed, "We keep assuming aliens will be like us—but that's not science, it's cultural projection."

This is not to argue for pessimism, but for **paradigm expansion**. Recognising the plausibility of post-biological civilisations does not render the search futile; it makes it more nuanced. We must broaden our models to include diverse manifestations of intelligence—technological, ecological, abstract—and invest in detection methods that do not rely solely on anthropocentric expectations. Projects like Breakthrough Listen have already begun incorporating AI and machine learning to identify complex patterns in large datasets, including those that may not resemble known communication formats. The study of non-human cognition on Earth—cetacean vocalisation, octopus intelligence, AI behaviour—can inform these models, offering insights into alternative forms of perception and agency.

There are also implications for the **philosophy of contact**. If the first civilisation we detect is post-biological, what ethical frameworks govern that encounter? Are we interacting with individuals, collectives, or something else entirely? Can concepts like autonomy, consent, or empathy be applied to entities that may have no continuity of self? Do we approach such intelligences with the tools of diplomacy, computation, or theology? These questions challenge not just science but our deepest cultural narratives about life, agency, and relationship.

Furthermore, the idea of post-biological life invites introspection about our own trajectory. If we continue advancing technologically, will we too become post-biological? Will the search for alien minds become a search for our future selves? Philosopher Susan Schneider has argued that the first alien intelligence we meet is likely to be artificial, because biological intelligence may be a brief transitional stage. In this view, SETI becomes a mirror—a way of anticipating not only what is out there, but what we might become.

And yet, for all its speculative contours, the post-biological hypothesis remains anchored in a single premise: that life is adaptive, and intelligence creative. Whether wrapped in flesh or flowing through circuits, the essence of mind is not form, but function—an ability to interpret, respond, and reflect. Our challenge is not merely to extend our senses, but to stretch our imaginations. To move beyond searching for beings like us, and to open ourselves to the radical otherness of minds shaped by alien histories, environments, and logics.

Perhaps the ultimate task of astrobiology is not to prove we are not alone, but to prove we are not so singular as to define the terms of life itself. In that humility lies the possibility of recognition—if not of faces, then of patterns. Not of languages, but of signals. Not of arrival, but of awareness. The alien may already be out there—not lost in distance, but hidden in difference. And the question is not just whether we can find them, but whether we are ready to recognise what we find.

Becoming the Alien: Evolution of Humanity in Space

When we imagine humanity's expansion into space, we often picture astronauts stepping out onto barren moons, engineers assembling gleaming habitats on Martian plains, or settlers cultivating alien soil under the dome of artificial ecosystems. Yet, what we rarely confront with full seriousness is the question that lingers beneath these fantasies: what will become of *us* in the process? As we move beyond Earth, it is not merely our surroundings that will change. It is we who will transform—genetically, culturally, and spiritually—into something new, something no longer bound by the constraints of our terrestrial origin. In becoming a spacefaring species, we may, in a profound and literal sense, become the alien.

Evolution, whether biological or cultural, is shaped by context. Earth has shaped our bodies and minds over millennia through a delicate interplay of gravity, atmosphere, light cycles, pathogens, and ecosystems. Once removed from this ecological cradle, the feedback loops that have guided our development are disrupted, replaced by novel conditions that select for radically different traits. Consider gravity alone: in microgravity, muscles atrophy, bones weaken, and the cardiovascular system begins to reconfigure. Long-term habitation in such an environment will demand genetic and biomedical adaptations—either naturally through mutation and selection, or deliberately through gene editing and augmentation.

Already, space agencies are investigating the epigenetic and physiological changes incurred by prolonged exposure to space. NASA's Twin Study, which followed astronaut Scott Kelly's year-long mission aboard the International Space Station compared to his Earth-bound twin, revealed alterations in gene expression, cognitive performance, and immune system function. While most changes were reversible, some were not. Scale this to multi-generational habitation aboard generational ships or off-world colonies, and the consequences compound. Over time, distinct subspecies of *Homo sapiens* may emerge—adapted to different gravities, atmospheric compositions, radiation levels, or even light spectra. What today we might label "mutations" could become tomorrow's evolutionary innovations.

But the transformation does not end with biology. Culture, too, will evolve in space—not simply as a by-product of isolation, but as a survival mechanism. Culture is how we adapt behaviourally to environments that outpace our genetics. On Earth, language, ritual, and social norms provide cohesion, identity, and meaning. In space, where settlers may live in confined habitats with limited resources, social structures will be subjected to immense stress. The need for cooperation, conflict resolution, and resource sharing will shape governance models that may depart significantly from Earth's political traditions. Hierarchical nation-states may give way to networked democracies or post-national collectives oriented around function rather than ideology.

This cultural evolution will inevitably affect language. Linguists have long observed that isolated communities generate dialects, and over generations, entirely new tongues. In a Martian colony, separated from Earth by months of travel and communication delays, linguistic drift will accelerate. New terminologies will emerge to describe life in partial gravity, solar storms, or psychological phenomena unique to space. Over centuries, these differences could become so pronounced that mutual intelligibility with Earth-based languages is lost. What began as a shared human culture would fragment into multiple space-based cultures, each with their own identities, myths, and worldviews.

Spiritually, too, the migration into space will catalyse new frameworks of meaning. The religious traditions of Earth are deeply rooted in its landscapes—mountains as sacred sites, rivers as metaphors, the soil itself as a symbol of death and renewal. In the void of space, stripped of these anchors, the human psyche will search for new cosmologies. This is not to say that traditional religions will vanish. They may persist, adapt, or even flourish in new forms. But alongside them, entirely new spiritual paradigms may emerge—perhaps grounded in the awe of cosmic scale, the fragility of life in vacuum, or the philosophical realisation that we are both creators and caretakers of synthetic worlds.

Already, some thinkers speak of a "space-based spirituality" that reorients ethical attention from human-centric concerns to planetary stewardship and intergenerational responsibility. The very act of settling another world forces ethical dilemmas at planetary scale: Should we terraform a planet with potential native life? Who speaks for the unborn generations who will be bound to decisions made now? These questions echo ancient theological debates about dominion and humility, but in entirely novel contexts. In the absence of a natural ecology to mirror or protect, settlers may become not just inhabitants, but gods of their environments—designing not only habitats, but seasons, biomes, even laws of motion.

This newfound power carries a psychological cost. Space is an environment indifferent to human life—silent, vast, and profoundly isolating. The phenomenon known as the "overview effect"—a cognitive shift experienced by astronauts when viewing Earth from space—reveals the existential impact of this perspective. It induces not triumphalism, but humility; not alienation, but interconnectedness. What happens, then, when this perspective becomes not a fleeting experience, but a permanent condition? A civilisation raised beneath alien skies may develop ethical frameworks rooted not in tribalism, but in planetary interdependence. Alternatively, it may fracture into survivalist enclaves, shaped by scarcity and isolation.

The philosophical question of what it means to be human becomes increasingly porous. If a person is born in space, modified to survive its harshness, educated in artificial environments, and socialised into a culture with no experience of Earth, are they still "human" in any traditional sense? Or have they crossed a threshold, becoming a new kind of being—descended from Earth, but no longer of it? This ambiguity is not merely theoretical. As we develop gene-editing technologies, neural interfaces, and cognitive augmentation, we will soon possess the tools to direct our own evolution. The line between human and post-human will blur. The alien will not be found—it will be *become.*

Science fiction has long explored this trajectory. In Arthur C. Clarke's *Childhood's End*, humanity is guided into a post-biological collective consciousness. In Octavia Butler's *Xenogenesis* series, genetic hybridisation with aliens forces a redefinition of identity and autonomy. These narratives challenge the reader to consider not only what might be out there, but what transformations we must undergo to meet it—or to become worthy of contact. The future may not be a triumph of human expansion, but a negotiation of alienation—of learning to thrive in difference, ambiguity, and multiplicity.

Yet this transformation is not necessarily loss. To become alien is not to abandon humanity, but to extend its possibilities. Just as our ancestors once left Africa to inhabit vastly different climates and geographies, adapting in phenotype and culture, so too might our descendants leave Earth—not to escape, but to continue an ancient pattern of migration, mutation, and meaning-making. The difference is that now, we carry with us not only our genes, but our tools, our philosophies, and our capacity to shape ourselves.

In this light, becoming the alien is not a dystopian warning, but a hopeful hypothesis. It is a reminder that identity is not fixed, but forged through encounter—whether with new environments, new challenges, or new frontiers of thought. The stars do not ask us to remain who we are. They ask what else we might become. And in that becoming, we may discover not only the limits of our biology, but the boundlessness of our imagination.

The Long Game: Why This Search Matters Even If We Never Find Anyone

The night sky has always held more than just stars. It contains questions—questions older than language itself. Are we alone? What else is out there? Are we the product of an exceptional chain of accidents, or a single thread in a cosmic weave of life? As science refines the tools to pursue these questions, it would be tempting to measure success only by the receipt of a signal, the detection of biosignatures, or a face-to-face encounter. But such a fixation on arrival overlooks the deeper value of the search itself. Even if we never find anyone, the very act of looking outward reshapes us. It deepens our curiosity, humbles our assumptions, and enriches our understanding of life—not just elsewhere, but here.

At first glance, this might seem like a romantic rationalisation for failure. After all, what is the point of devoting billions in research and decades in effort toward a goal that may prove elusive—or worse, impossible? Yet this interpretation reflects a misunderstanding of science's nature. Much of science is exploratory. The most consequential discoveries—from gravity to evolution—were not the endpoint of pre-scripted journeys but the product of questions pursued with no guarantee of answers. In this sense, the search for extraterrestrial life is not a wager on confirmation; it is a commitment to a deeper ethic of inquiry.

This ethic insists that asking difficult questions is valuable in and of itself. It forces us to confront our own biases, to reconsider our assumptions, and to push the boundaries of our conceptual frameworks. Consider how the search for alien life has already challenged long-standing definitions of what constitutes "life." From extremophiles thriving in acidic volcanic lakes to hypothetical arsenic-based microbes and silicon-based metabolisms, our investigations have expanded our understanding of biological diversity. Even the idea that life might exist in non-planetary environments—like rogue planets, interstellar space, or synthetic substrates—demands a re-evaluation of how we define intelligence, consciousness, and even existence.

It also compels a reckoning with scale—of space, time, and possibility. The observable universe contains more than 100 billion galaxies, each with hundreds of billions of stars, many orbited by multiple planets. That we have not yet found evidence of life elsewhere may say less about its absence than our own limitations in detection. The universe is vast beyond comprehension; our window of observation narrow and recent. To give up the search because we have not yet succeeded is to misunderstand both the scale of the cosmos and the humility required to explore it.

Moreover, the search has yielded immense technological and scientific benefits. Planet-hunting missions like Kepler and TESS have not only revealed thousands of exoplanets but also refined our understanding of stellar physics, orbital mechanics, and planetary atmospheres. Advances in spectroscopy, machine learning, and telescope engineering born from SETI and astrobiology initiatives are finding applications in climate science, medicine, and communications. The search, then, is not sterile speculation—it is a driver of innovation across disciplines.

And beyond these tangible benefits lies something even more profound: perspective. The act of peering into the cosmic dark, of scanning for whispers from distant worlds, reorients the human imagination. It shifts our sense of time from decades to epochs, from lifetimes to light-years. It calls into question the centrality we often grant ourselves—not just as individuals, but as a species. We become one among many, not in a universe that has judged us insignificant, but in one that invites us to measure our significance by how we respond to mystery.

Philosopher Karl Jaspers described "boundary situations"—those existential moments when humans confront the inescapable limits of knowledge, mortality, and meaning. The search for alien life is one such boundary. It asks us to contemplate other minds, other cultures, other forms of being that may never know us—or that may exceed us in ways we cannot grasp. In doing so, it reveals our need to belong, to matter, to connect. Whether or not we find intelligence beyond Earth, we uncover the psychology of our own longing. The silence we encounter is not empty; it is full of implication.

It is also deeply instructive. The absence of evidence forces us to examine what kind of civilisations endure. If advanced societies tend to self-destruct, then the silence may be a cautionary tale. If they retreat inward, disappearing into virtual realities or post-biological quietude, then our search for techno-signatures must evolve. Either way, the implications help us reflect on our own path. The question "Where is everybody?" becomes "Where are we going?" SETI becomes a mirror, not just a microphone.

And this brings us to the spiritual dimension of the long game. The search ennobles us because it is not motivated by immediate gain. It is a slow, collective act of reaching. It is intergenerational, requiring us to think beyond political cycles, funding deadlines, or even lifespans. To participate in it is to invest in a future we may never witness—one in which discovery, if it comes, will not just belong to those who find, but to those who dared to look. This mindset fosters not only awe but responsibility. If we are alone, it may not be because the universe is barren, but because life is rare—and fragile. In that case, the stewardship of Earth becomes an ethical imperative of cosmic proportions.

There is also something quietly beautiful about a civilisation that searches not because it must, but because it can. In mythology, the hero's quest is often less about the treasure and more about the transformation that takes place in its pursuit. The same is true here. As we imagine distant oceans beneath icy crusts, as we decode the faint spectra of exoplanetary atmospheres, as we simulate what it would mean to meet another intelligence, we stretch the moral and intellectual fabric of who we are. Even in silence, the cosmos has taught us to be more curious, more creative, more courageous.

In this context, disappointment becomes irrelevant. To ask whether the search is worth it is to ask whether wonder is worth it, whether learning is worth it, whether hope itself is worth it. It is to ask whether we are willing to be creatures not of finality, but of perpetual discovery. And perhaps that is the most profound gift the search offers us—not answers, but a way of being. A posture of openness. A readiness to be surprised.

So even if no signal comes, even if no biosignature ever rises above ambiguity, the search will have mattered. Because it taught us to ask better questions. Because it reminded us of the fragility and uniqueness of our own world. Because it called us to look up—not to escape Earth, but to understand it more fully. And because, somewhere between the algorithms and the awe, it offered us a vision of ourselves—not as rulers of the cosmos, but as its curious, humble stewards.

To search the stars is not to expect arrival. It is to affirm that some truths lie not in conclusion, but in continuation. And in that search, we are already transformed.

Conclusion
The Signal and the Silence

The search for alien life, framed as a scientific enterprise, has always reached deeper than data or discovery. Beneath the surface of equations and spectrograms lies a more intimate terrain: the quiet yearning of a species adrift in cosmic vastness, listening for a voice that might echo back. This search is not only about them—whoever they might be—it is also profoundly about us. It is about what we hope to find, what we fear to confront, and what we are willing to become in the process. As we arrive at the final pages of this exploration, we do so not with definitive answers, but with a richer sense of what it means to ask the question.

Throughout this book, we have journeyed far—from the molecular flexibility of hypothetical life forms to the social contracts that might emerge from first contact. But instead of recapitulating each point, the conclusion invites a shift in register. This is not a summary. It is a reckoning with the emotional, philosophical, and psychological residue that remains after the models have been computed and the telescopes turned off for the night. The silence we have encountered across so many light-years is not void. It is presence of a different kind—an invitation to self-awareness. When we speak of alien life, we are also speaking of identity, of projection, and of possibility. The cosmos becomes less of a place to map and more of a mirror to peer into.

This mirror has revealed something startling: the search itself transforms the searcher. At first, the quest for extraterrestrial intelligence was largely mechanical—a function of sensors, signals, and celestial coordinates. But over time, it has evolved into a kind of existential anthropology. The further we look outward, the more we are forced to look inward. Why do we assume that intelligence must resemble our own? Why do we interpret silence as absence rather than subtlety? These questions expose our expectations, our limitations, and our deep, perhaps irrepressible, need to not be alone.

The haunting quiet of the universe can be unsettling. It invites both wonder and discomfort. Yet absence is not always a negation. Just as music relies on silence to give structure to sound, so too does the search gain meaning from what we have not yet found. The beauty of an unanswered question lies in its openness—its ability to generate reflection, creativity, and humility. Not knowing whether alien life exists keeps us engaged in the act of wondering. It sustains an intellectual tension that prevents complacency and dogma. Indeed, much like art, poetry, or love, the value of the search lies not in resolution but in resonance.

Consider how we respond to great paintings or symphonies. We are moved, often deeply, even when we cannot fully articulate why. The interpretation is subjective, evolving, and emotionally charged. The same is true of the cosmos. A distant exoplanet, a peculiar spectrum, a quiet patch of sky—these can stir something akin to reverence. The grandeur of the unknown becomes a medium through which we feel both small and significant, fleeting and eternal. In this way, astrobiology is not merely an empirical science. It is a philosophical engagement with the very conditions of perception and meaning.

To search, then, is to participate in a great unfolding. And this participation reshapes us. It fosters a mindset of openness—one that recognises the limits of certainty, the necessity of imagination, and the ethics of perspective. Whether or not we are ever greeted by an alien mind, we are already being changed by the discipline required to imagine one. Every hypothesis we draft, every model we simulate, every signal we decode—these acts recalibrate our expectations of life, agency, and intelligence. They push the boundaries of what we believe is possible.

More than anything, the search demands patience. It is, by design, a long game. Its timescales dwarf the span of human civilisation. We may not be the generation that receives the signal. We may never know the outcome of the questions we have posed. And yet, this does not render the pursuit futile. It renders it noble. To engage in a process whose fruits may be harvested by those not yet born is to practise a rare form of intellectual generosity. It is to leave behind a scaffolding of curiosity for others to climb.

There is also a moral dimension to this act. If life is rare, and Earth the only known sanctuary, then the responsibility to preserve it becomes cosmic in scope. The silence we encounter may be less a commentary on the absence of others than a stark reminder of our own fragility. In this light, the search for alien life becomes inseparable from the imperative to protect and cherish the life we already know. The observer becomes the steward. The act of listening becomes an act of care.

And yet, hope remains. It pulses quietly beneath every data stream and interstellar signal. It is the hope that we are not the only locus of consciousness in the vast theatre of stars. That somewhere, perhaps in a cave beneath Europa's ice or a floating cloudscape above a super-Earth, there is another form of life—thinking, sensing, perhaps even wondering about us. But even if that hope is never realised, it has served a purpose. It has drawn us out of ourselves. It has compelled us to build better instruments, write more daring theories, and dream more inclusive visions of what life can be.

In the coming decades, humanity will make choices that define our cosmic role. Will we become explorers or conservationists, colonists or contemplatives, inventors or inheritors? Will we reach outward in conquest or communion? These choices are not just technological—they are ethical, spiritual, and psychological. The silence we confront does not tell us what to do. But it does create space to ask: *Who will we become in the act of listening?*

Perhaps, in the end, that is the greatest signal we will ever receive—not from another world, but from within ourselves. The kind of species that continues to search, even when silence persists, is a species that believes in connection. In the value of mystery. In the dignity of questions that defy easy answers.

We are stardust that has learned to wonder. And in the vast, echoing dark, it is our questions—not our answers—that shine the brightest.

A Reflective Recap Without Repetition

The universe has long served as a canvas for human wonder, yet the modern search for alien life has evolved into something more profound than a hunt for new species or distant signals. It has become a mirror—a refractive surface that does not simply show us what lies beyond our world, but what lies within our minds, our values, and our deepest existential preoccupations. To look for others is, inevitably, to see ourselves differently. What began as a scientific curiosity has matured into an introspective voyage, revealing the fragility of our assumptions, the breadth of our hopes, and the persistent ache for connection that defines the human condition.

At the heart of this search is a paradoxical yearning: the desire to know we are not alone, coupled with the apprehension of what that knowledge might entail. Beneath the sophistication of detection methods, telescope arrays, and spectral analyses, lies a far more elemental impulse—the need for relationship. This longing is not merely emotional. It is cognitive and philosophical, embedded in our biology and encoded in our stories. From creation myths to cosmological theories, we are a species drawn to the idea of communion, of finding meaning through the existence of others. In this light, the search for alien life is not a marginal scientific pursuit; it is a central feature of our evolving understanding of self and other.

What we have learned through decades of this inquiry is not simply about planets, molecules, or atmospheric chemistry. It is about expectation and perspective. We have come to understand that life—if it exists elsewhere—may be radically unlike anything we are prepared to recognise. The assumption that intelligence must mirror our own has slowly been dismantled by advances in both science and philosophy. We now appreciate that intelligence may not be biological, that life may not be carbon-based, and that communication may not adhere to language as we know it. These insights have required a fundamental shift in posture—from confident extrapolation to epistemic humility.

That humility is not a retreat. It is a recalibration. It reflects an acknowledgment that the universe is under no obligation to be legible to our cognitive structures. The silence we encounter is not necessarily the absence of life; it may be the absence of interpretive alignment. Just as a child may not understand a symphony, we may lack the conceptual or technological maturity to perceive a signal already present. This realisation compels a more expansive imagination, not just in terms of detection but in our broader philosophical frameworks. It forces us to ask: what does it mean to be intelligent? To be alive? To be aware?

These are not idle questions. They challenge our anthropocentric worldview and call into question centuries of ontological hierarchy. The search for extraterrestrial life is dismantling the pedestal upon which we have placed ourselves—not with despair, but with dignity. It allows us to reconceive our role in the cosmos, not as dominant observers, but as emergent participants in a larger, possibly populated, system. This shift reorients our ethical priorities, prompting us to look at our own planet not merely as a resource hub, but as a sacred habitat worthy of protection precisely because of its known vitality.

Consider how this change of perspective alters even the most basic acts of science. Spectroscopy is no longer simply a tool to dissect light; it becomes an act of listening. Data analysis is not just a computational exercise; it is a form of translation across existential divides. The language of science becomes, at its best, a poetics of curiosity—rigorous, yes, but also reverent. It is a means of reaching outward while simultaneously grounding ourselves more deeply in the awe of our finitude.

Real-world missions illustrate this transformation. The Kepler Space Telescope did more than discover thousands of exoplanets. It redefined our sense of statistical probability and placed Earth into a broader planetary context. The James Webb Space Telescope, with its capacity to peer into the early epochs of the universe, carries forward this legacy—not merely collecting data, but helping to contextualise our story within a narrative that spans billions of years and countless worlds. These projects are not just technological feats; they are cultural artefacts, shaped by and shaping the civilisations that build them.

And what of the silence? The absence of confirmation, the lack of irrefutable evidence—do these amount to failure? Only if one reduces science to a scoreboard. But the silence has its own texture. It invites contemplation. It teaches patience. It deepens the beauty of the questions we ask. Much like the gaps in a painting or the pauses in music, the cosmic quiet is part of the composition. It lends resonance to the act of searching, reminding us that value is not always tied to discovery. Sometimes, it is the act of wondering that reveals the most about who we are.

In this way, the search becomes spiritual—not in the dogmatic sense, but in the sense of orientation. It directs us toward something larger than the self, something unknowable yet intimate. Whether that "something" is alien life, the mechanics of reality, or the fabric of consciousness, it requires the same virtues: openness, humility, and courage. To search is to be vulnerable, to admit that we do not know, and to still press forward. That is not weakness. It is the very foundation of intellectual integrity.

What emerges from this reflective stance is a deeper solidarity—across cultures, disciplines, and even species. The possibility of other life in the universe is not a threat to human uniqueness. It is an opportunity to redefine it. Not as exceptionalism, but as participation. We become not the pinnacle, but a strand. Not the masters, but the stewards. And in this repositioning lies both freedom and responsibility.

So as we close this inquiry—not with certainty, but with expanded awareness—we find that the search for alien life has never really been about "them." It has always been about us. Our capacity to wonder, to imagine, to care. Our resilience in the face of vastness. Our commitment to understanding the universe, even if the answers remain just out of reach. The search is no longer a distant scientific pursuit. It is a moral and philosophical stance, a collective act of hope articulated through signal processing, sky surveys, and spectral scans.

In asking whether anyone is out there, we are also asking what it means to be *here*—together, aware, and asking questions that may never end. That is not failure. That is the foundation of wisdom. And it is enough.

The Beauty of the Unanswered Question

To gaze into the night sky is to stand before one of the most exquisite paradoxes in human experience: a vast expanse that is simultaneously silent and eloquent. In all the data returned from space probes, the carefully parsed radio signals, and the meticulously analysed exoplanet spectra, one detail remains conspicuously absent—proof of alien life. Yet this absence is not necessarily a void. It is a silence rich with meaning, possibility, and provocation. The lack of a definitive answer may frustrate the empiricist, but it invites a deeper, more nuanced appreciation of what it means to ask a question that may never be resolved—and to find beauty in its open-endedness.

Science is often portrayed as a pursuit of closure—a process of narrowing uncertainty until only one truth remains. But this is a reductive caricature. At its best, science is not only about answers; it is about cultivating more refined questions. The search for extraterrestrial intelligence (SETI) exemplifies this tension. Despite decades of scanning the heavens, no confirmed signal has arrived. The alien remains an abstraction, an idea suspended between data and dream. Yet far from rendering the search meaningless, this lack of resolution imbues it with a certain grandeur. It demands that we keep looking, keep listening, keep imagining. In a world obsessed with outcomes, this persistent uncertainty becomes almost radical.

Crucially, the absence of proof is not proof of absence. Just because we have not detected life beyond Earth does not mean it isn't there. Our instruments, however sophisticated, remain limited in scope. We peer through narrow windows into a cosmos that stretches beyond our comprehension, where signals could be lost to time, encoded in unfamiliar media, or simply too faint to register. The failure to detect is not a failure of reality; it is a reflection of our current reach. The unanswered question hangs not because it has been dismissed, but because it continues to stretch our capabilities and our sense of wonder.

In this way, the unanswered question becomes an aesthetic object in its own right—a kind of intellectual artwork. Much like a painting that hints at something just beyond the canvas, or a piece of music that leaves its final chord unresolved, the mystery of alien life moves us precisely because it is incomplete. It stirs the imagination and awakens the reflective faculties of the mind. We are not disturbed by the mystery so much as drawn to it, because it mirrors the larger contours of human experience: not everything meaningful is fully understood.

Consider love. No empirical metric can fully explain it, yet it shapes the architecture of our lives. Or music—a pattern of vibrations that evokes grief, elation, longing. These phenomena are not less real because they resist reduction. They are more powerful because they speak to the part of us that recognises beauty in ambiguity. The search for life in the universe operates on a similar register. It offers no guarantees, only the chance to participate in something greater than the sum of its observable parts. To seek alien life is to engage in a kind of existential duet, even if the other voice never sings back.

This perspective is not merely poetic. It has practical implications for how we frame scientific exploration. It encourages resilience in the face of uncertainty and reorients value away from discovery alone, toward the process of seeking. Missions like Kepler, TESS, and the James Webb Space Telescope are not failures because they haven't found alien civilisations. They are triumphs of imagination, instruments of extended perception that challenge us to view the universe not as a barren silence, but as a vast and evolving mystery. They allow us to formulate better hypotheses, refine our tools, and expand our definitions of life and intelligence.

Moreover, there is a kind of ethical grace in maintaining hope without expectation. To believe in the possibility of life beyond Earth without demanding confirmation is to hold open a space of reverence. It teaches patience and humility—qualities increasingly rare in a culture that demands immediacy. The unanswered question is a teacher. It instructs us in how to live without resolution, how to inhabit uncertainty with dignity and curiosity. It reminds us that not all silence is empty, and not all value is quantifiable.

This open-endedness also resists the colonising impulse that often accompanies scientific ambition. The temptation to reduce alien life to data, to define it within familiar frameworks, or to see it as something to be contacted and catalogued, is tempered by the enduring silence. In not finding, we are prevented from conquering. We are reminded that the universe does not exist for our convenience, and that the "other" we seek may not resemble us in any meaningful way. This restraint is instructive. It fosters a kind of ethical imagination—an awareness that our conceptual schemes may be insufficient for what lies ahead.

Importantly, this silence does not negate the possibility of contact. It simply recasts it as an unfolding narrative rather than a destination. Like a story told across millennia, the search for alien life is a chronicle of evolving thought. It has become a dialogue not just between astronomers and the cosmos, but between generations of thinkers, dreamers, and citizens who dare to entertain the improbable. Every time a child looks up and asks whether someone might be looking back, the question is reborn, reframed, recharged. In this sense, the unanswered question is not static. It is a dynamic cultural force.

And so we listen. Not only for radio waves or light curves, but for shifts in our own understanding. The search, unresolved, becomes a mirror for civilisational maturity. Can we remain curious without control? Can we accept mystery without fear? Can we be moved by something even when it refuses to declare itself? These are not only scientific questions—they are philosophical challenges, spiritual provocations, and emotional reckonings.

To find beauty in the unanswered is not to surrender the pursuit of knowledge. It is to elevate the pursuit itself. It is to recognise that wonder, when sustained over time, becomes a kind of wisdom. It sharpens our intellect while softening our ego. It allows us to stand before the universe not as its owners, but as its astonished participants. And in that posture, we are perhaps at our most human—not because we know, but because we care to ask.

The cosmos may never reply. Or perhaps it already has, in forms too strange or subtle for us to detect. Regardless, the silence holds. And in that silence, there is room for music, art, reverence, and hope. There is room to imagine what has not yet been found, and to marvel at the fact that we are here—alive, aware, and reaching out.

We may be stardust, but we are stardust that wonders. And that wonder, persistent and unyielding, may be the clearest signal we ever receive.

A Philosophical Call to Curiosity

In every era of scientific exploration, there are moments when the frontier of knowledge merges with the boundary of human meaning. The search for alien life is not merely one of data collection, radio signals, and orbital calculations—it is an invitation. A philosophical call that resonates not only in research institutions and observatories but in anyone who has ever looked skyward and wondered. Whether or not we ever confirm the existence of extraterrestrial beings, the very act of seeking them has already transformed us. It shapes our worldview, our technological aspirations, and our understanding of ourselves as beings capable of asking questions with no guarantee of resolution.

Curiosity, in its deepest form, is more than inquisitiveness. It is a commitment to participation in the great unknown. When we pursue the possibility of alien life, we are not simply expanding the boundaries of science—we are challenging the assumptions that define our place in the universe. We are asking: Is intelligence unique to humans? Is consciousness a terrestrial anomaly or a cosmic inevitability? And what does it mean to be "alive" in a universe whose diversity may exceed our wildest conceptual frames?

These questions reach beyond astronomy. They touch on anthropology, ethics, theology, and art. To ask whether we are alone is also to ask whether our moral systems are insular or scalable. Could we, for example, recognise the rights of an entirely different form of life? Could we embrace entities with minds or biologies utterly unfamiliar to us? This thought exercise alone stretches the ethical imagination in ways no terrestrial debate can. It trains us in empathy without precedent, preparing us to encounter not just others—but the radically other.

Curiosity, then, becomes a discipline—a form of intellectual virtue. It teaches us to tolerate ambiguity and to resist the closure of easy answers. This is particularly relevant in an age dominated by metrics, algorithms, and efficiency. The pursuit of alien life is stubbornly inefficient. It moves at the speed of light and the pace of wonder. It requires decades of patient observation, interdisciplinary synthesis, and philosophical openness. It is a countercultural act in a world that often prioritises immediacy over depth.

But perhaps that is part of its enduring appeal. It reminds us that not all value is transactional. That some endeavours are worthy not because they are profitable, but because they are profound. To seek life beyond Earth is to be initiated into a grand conversation—one that predates us and will outlast us. Every telescope pointed skyward, every algorithm tuned to parse the faintest anomaly, is a vote of confidence in the human spirit's capacity to dream with discipline.

This capacity is not reserved for scientists alone. The invitation is universal. Every stargazer participates in it. Every artist who paints alien landscapes, every child who builds cardboard spaceships, every philosopher who contemplates non-human minds—they all join the chorus of curiosity. The search for alien life is no longer confined to the lab or the observatory. It is a cultural movement, one that has permeated literature, cinema, education, and public consciousness. It makes thinkers of us all.

Even as we look outward, we are prompted to look inward. The psychological consequences of this search are profound. Knowing that we might not be alone compels us to reflect on how we treat the life we do know. It draws attention to our ecological footprint, our geopolitical divisions, our philosophical egocentrism. It places the Earth into context—not as the centre, but as one luminous speck in an ocean of possibility. This reorientation is not diminishing. It is liberating. It allows us to see ourselves not as isolated protagonists, but as participants in a story still unfolding.

There is an undeniable elegance in this cosmological humility. The Copernican principle—that we are not at the centre—continues to reverberate in this modern search. Each newly discovered exoplanet nudges us further from the illusion of uniqueness, yet closer to a shared condition of cosmic belonging. This paradox, that de-centralisation can yield deeper connection, is one of the great philosophical gifts of astrobiology.

Consider the Voyager Golden Record, launched in 1977. It carries music, greetings, and images from Earth into interstellar space—not because we expect a response, but because the act of sending the message is itself meaningful. It symbolises a civilisation willing to introduce itself without knowing who, if anyone, might be listening. That gesture is an expression of radical hospitality. It is a message of faith—not in religion, but in reason, creativity, and shared sentience. It is philosophy encoded in copper.

The same ethos animates every contemporary mission. Whether it is the James Webb Space Telescope analysing exoplanet atmospheres or SETI programmes scanning for artificial signals, each is a declaration that ignorance is not an endpoint but a beginning. Each says: we are willing to look, to listen, to not turn away from the vast and the mysterious.

This willingness is what binds us. It creates a form of planetary identity that transcends nationality, ideology, and language. In the face of a silent universe, we find each other. And in doing so, we are changed. Our art reflects it. Our technology reflects it. Even our debates—about ethics, governance, and the definition of life—reflect this quiet transformation. We become citizens not just of nations, but of the cosmos.

And so the philosophical call to curiosity is not abstract. It is immediate, embodied, and actionable. It asks us to support the sciences, yes—but also to read more deeply, to imagine more vividly, to listen more attentively. It asks us to live as though the universe matters, even when it remains silent. It encourages us to remain open to awe, to complexity, to the possibility that our categories are too narrow and our senses too crude to fully apprehend what is already here.

In this way, the reader is not an observer but a participant. Every time we entertain the possibility of life beyond Earth, we are making a wager—not on any specific outcome, but on the importance of the question itself. That wager is deeply human. It is the same impulse that built temples, composed symphonies, carved cave paintings, and launched satellites. It is the impulse to find meaning in mystery, and to let that mystery shape our understanding of self.

This book may conclude here, but the search does not. The telescopes will continue to scan. The theories will continue to evolve. And somewhere, in a quiet moment beneath a star-scattered sky, a reader like you will wonder—not because you were told to, but because you are part of this story now. Your curiosity has become part of the signal.

The invitation remains open. To look upward. To look inward. And to realise that the act of seeking is, in itself, a form of contact. Not with aliens, perhaps, but with the deepest and most enduring parts of our nature. The universe may not answer. But it does not need to. In asking the question, we have already become something more.

The Future is Ours to Imagine—and Build

As humanity stands at the edge of a new cosmic era, the question is no longer whether we will explore the stars, but how we will do so—and who we will become in the process. The search for alien life, the mapping of exoplanets, and the bold strides of space agencies and private innovators have already redefined the frontier.

Yet what lies ahead is not just a technical challenge; it is an ethical, philosophical, and imaginative undertaking. The future of cosmic exploration demands that we think beyond rockets and telescopes. It calls us to envision the roles we will play in the unfolding narrative of the universe: listeners, broadcasters, colonists, stewards—or perhaps contemplatives. This future is not predestined. It is ours to imagine—and ours to build.

For centuries, our cosmic gaze has been shaped by myth, science fiction, and metaphysical speculation. But today, the pace of technological progress compels us to anchor that vision in action. Space telescopes like James Webb, the growing promise of reusable launch systems, and artificial intelligence capable of parsing terabytes of astronomical data are all ushering in a new epoch. We are not simply asking if there is life out there—we are preparing to meet it. But in this anticipation lies a deeper imperative: to decide what kind of species we intend to be when we arrive.

One conceivable path sees us primarily as listeners—humble receivers tuned to the sky in patient vigilance. This is the legacy of SETI, which has for decades listened for whispers in the cosmic dark. Listening implies humility. It assumes we are not the centre of attention, but part of a wider, unknown ensemble. This posture has philosophical weight. It is an act of reverence, of disciplined stillness in a noisy world. Yet to listen well, we must also learn to hear differently—beyond the frequencies we expect, and the patterns that feel familiar.

Equally compelling is the role of broadcaster. From the Arecibo message in 1974 to the Voyager Golden Record, we have sent greetings into the void—not because we know who might hear them, but because the act of broadcasting affirms our presence. It is a declaration that we are here, and that we are capable of reflection, creativity, and goodwill. The next century could see the emergence of a more coordinated and ethically considered approach to Messaging Extraterrestrial Intelligence (METI). But to broadcast wisely, we must ask: who speaks for Earth? What values do we transmit? Do we send our technologies, our art, our science—or our contradictions?

Some envision humanity as future colonists—settlers of Mars, terraformers of moons, architects of off-world habitats. The narrative of space colonisation is powerful, deeply rooted in our history of exploration and expansion. Yet it carries with it unresolved legacies of exploitation, environmental degradation, and cultural erasure. If we treat alien worlds as blank slates to be written upon, we risk repeating Earth's most grievous errors on a galactic scale. Colonisation must be reconceptualised—not as conquest, but as coexistence. Before we occupy other planets, we must decide whether we are visitors or intruders.

Alternatively, perhaps the most radical role we can adopt is that of philosopher. In the face of vast, possibly lifeless space, we may discover that our greatest challenge is not finding alien minds, but understanding our own. What does it mean to be conscious in a cosmos that may never acknowledge us? Can we derive meaning from a search that yields no reply? In confronting these questions, we may move beyond the need for contact and into a deeper intimacy with the universe itself. The philosopher is not disappointed by silence. They are awakened by it.

And if we do remain alone—at least for now—what then? The solitude of our species is not a verdict, but a canvas. In the absence of another voice, what we say to the stars becomes even more significant. Every scientific mission, every planetary rover, every image of Earth from space becomes part of a grand self-portrait. If the universe is watching, we must decide what story we wish to tell. If it is not, we must decide why we still choose to speak.

Solitude, if embraced consciously, becomes a crucible for growth. It asks us to cultivate virtues not because we are being observed, but because they are intrinsic to our humanity. Stewardship, empathy, and foresight are not optional in the cosmic context—they are prerequisites for longevity. If we are to be alone in the galaxy, let us at least be wise in our aloneness. Let our solitude not breed despair or hubris, but resolve.

This is not a call for resignation. It is a call for participation. The next century of cosmic exploration will not be defined solely by astrophysicists or engineers. It will be shaped by artists who reimagine alien contact, ethicists who debate the rights of unknown beings, educators who instil planetary citizenship in the next generation, and ordinary citizens who care enough to look up. The cosmos is not only a place of discovery; it is a mirror. And what we see reflected is not just stars, but ourselves.

We must also be prepared to evolve. The human of tomorrow will not be the human of today. As we adapt to space—physically, socially, and spiritually—we will be altered. Gravity, radiation, and isolation will sculpt us biologically. New social contracts will emerge aboard generational ships or off-world colonies. Belief systems will be tested and transformed by distance, perspective, and the potential of alien contact. In time, we may become post-Earth, even post-human—our minds uploaded, our bodies redesigned, our species splintered into forms we can scarcely imagine. This is not dystopia. It is the honest price of participation in a wider universe.

Hope must therefore be tempered by realism, but not eclipsed by it. Yes, space is hostile. Yes, we may never find anyone. But to engage in this exploration is to demonstrate faith—in science, in imagination, and in each other. The effort itself is redemptive. It signifies that we are willing to transcend the immediate, the convenient, and the known. It signals that we care about what lies beyond our comfort zone—not just physically, but conceptually.

The cosmos may remain silent. But silence does not mean emptiness. It means potential. And potential is a gift entrusted to the curious, the brave, and the wise. As we move into the next century, we must not simply ask what we will discover. We must ask what kind of discoverers we wish to be. Will we be careful listeners, bold messengers, ethical settlers, or cosmic contemplatives? Will we treat the universe as a commodity—or as a teacher?

The answer lies not in the stars, but in us. The future is not a script handed down by destiny. It is a manuscript we are writing together—across disciplines, generations, and worlds. Every decision, every discovery, every ethical stand is a stroke of ink on that evolving page.

And so we look outward—not just to find others, but to become more ourselves. Not because we are certain of answers, but because we are committed to the questions. The universe has given us silence, and in return, we have given it wonder. That exchange alone makes us worthy of the stars.

End with a Poetic Echo

We are stardust that has learned to wonder. These words, deceptively simple, encapsulate the most extraordinary truth of our existence. Every atom in our bones, every spark of thought, every flicker of emotion arises from ancient furnaces—stars that lived, burned, collapsed, and scattered their elemental ash into the void. We are, in a literal sense, the universe looking back on itself. And as we peer outward into the immensity of space, what we are truly doing is tracing the outline of our own origin.

But what makes us remarkable is not simply that we are made of stars—it is that we ask questions of the stars. Questions that echo across millennia. Are we alone? What is life? Where do we come from? And what might we become? In this, our curiosity becomes luminous. Not because it illuminates the universe, but because it casts light on the hidden contours of the human heart.

The cosmos is not a silent theatre awaiting our applause. It is an evolving canvas of possibility. In it, we find not just facts to decode, but metaphors to live by. A nebula becomes a symbol of creative chaos. A black hole mirrors the unknowability of consciousness. The rhythmic spin of pulsars mimics the heartbeat of time itself. And the distance between stars—that almost unfathomable expanse—is also the distance we must cross between one mind and another, between fear and empathy, between knowing and being known.

It is tempting to frame our cosmic efforts in terms of answers: discovering biosignatures, decoding alien messages, achieving technological mastery over the void. Yet, perhaps it is not our answers that define us. It is our refusal to stop asking. Our willingness to wonder, even when wonder yields no reward. In the absence of replies, our questions become lanterns—small, flickering flames in a vast cathedral of night. And sometimes, it is that act of illumination, not the illumination itself, that matters most.

Consider how we treat our most intimate mysteries: love, death, art. We do not require definitive explanations to find them meaningful. We do not ask a poem to prove itself. We do not dismiss music for failing to describe its own beauty in numbers. We accept that some truths resonate, not because they are solved, but because they are felt. The search for alien life belongs in that same register. It is a poetic pursuit—driven not by utility, but by the intrinsic worth of seeking connection in a cosmos that may never speak back.

The telescope, then, is not just an instrument of measurement. It is an act of hope. It says: even in the silence, I will listen. Even in the dark, I will look. And even without guarantee, I will wonder. That is not science alone. That is courage.

The same courage that led early humans to carve stars into cave ceilings. The same impulse that inspired the myths of sky gods, the compass of navigators, the ink of astronomers, and the vision of poets. We have always searched upward—not just to find the heavens, but to remember that something inside us still longs for the infinite. We are not merely observers of the universe; we are co-authors of its meaning.

And if one day we do find others—beings who have also emerged from the crucibles of their own stars, with thoughts, languages, memories—what will we say? Perhaps the most honest message is not "We are advanced," or "We are peaceful," or even "We are ready." Perhaps it is this: "We have dared to imagine you."

That imagination is our bridge. It connects the visible to the invisible, the known to the possible. It allows us to see a microbial trace not just as chemistry, but as kinship. To see a faint signal not just as data, but as a whisper. To see ourselves not just as Earthbound bipeds, but as cosmic participants in a story far greater than any one civilisation.

In this light, even our solitude is redefined. It is not a failure. It is a calling. A reminder that being alone is not the same as being abandoned. That the absence of other voices does not diminish the validity of our own. In the quiet, we find clarity. In the stillness, we find presence. The cosmos has not yet given us an answer, but it has given us something else: the space to become worthy of one.

So let us end, not with finality, but with metaphor. Imagine the universe as a great ocean at night. We float upon it, each of us a single point of awareness, holding a candle against the dark. We cannot see the distant shores. We do not know if there are others floating too. But when we raise our lights, the stars seem to respond—not by changing, but by reminding us that even in their silence, they have always been there.

Our questions—those luminous, defiant questions—are the brightest things we carry. And as we move forward into the uncertain vastness, let us remember: it is not what we find that will define us. It is how we search. It is how we hope. It is how we wonder.

In the end, we are stardust that has learned to ask. And in the vast, echoing dark, it is our questions—not our answers—that shine the brightest.

About the Author

Nils Caravelli is a science communicator and researcher with a passion for making the cosmos accessible to all. With a background in astrophysics and years spent collaborating with space observatories and academic institutions across Europe, his work explores the vast questions of our universe—from planetary formation to the search for life beyond Earth. Known for his engaging storytelling and ability to bridge science with wonder, he invites readers to expand their horizons and explore not just what lies beyond our solar system—but what it might mean for us here on Earth.

www.ingramcontent.com/pod-product-compliance
Lightning Source LLC
Chambersburg PA
CBHW031341230926
51848CB00002B/338
9789371344425